W. Harcourt Bath

## The Young Collector's Handbook

Of Ants, Bees, Dragon Flies, Earwigs, Crickets, and Flies

W. Harcourt Bath

**The Young Collector's Handbook**
*Of Ants, Bees, Dragon Flies, Earwigs, Crickets, and Flies*

ISBN/EAN: 9783337273071

Printed in Europe, USA, Canada, Australia, Japan

Cover: Foto ©berggeist007 / pixelio.de

More available books at **www.hansebooks.com**

# THE YOUNG COLLECTOR'S
# HANDBOOK

OF

# Ants, Bees, Dragon-Flies, Earwigs, Crickets, and Flies

*(HYMENOPTERA, NEUROPTERA, ORTHOPTERA, HEMIPTERA, DIPTERA).*

BY

## W. HARCOURT BATH.

SECOND EDITION.

LONDON:
SWAN SONNENSCHEIN & CO.,
PATERNOSTER SQUARE.
1890.

# PREFACE.

MOST boys have a fondness for forming collections of various objects, such as Foreign Stamps, Crests, and Coins ; but very few comparatively collect Natural Objects. Now it will be admitted by all that the collecting of Natural Objects, such as Insects, Shells, Plants, Fossils, Minerals, etc., possesses immense advantages over that of Foreign Stamps and the like ; for the former, besides satisfying the collecting ambitions, also cultivates the observant and intellectual faculties, while at the same time affording healthful recreation in the fields and woods.

Again, a mere collector's province may be exhausted in a few years, whereas the study and observation of Natural History, which are the usual "fruits" of collecting Natural Objects, are practically inexhaustible.

Another great advantage which Natural History possesses, is that it may be prosecuted with very little expense, and is therefore a subject which even the poorest person may conveniently undertake.

This little handbook is intended to be a "Guide to Collecting Insects," which by way of Natural History, we may remark, is becoming more popular every day in this country. Hitherto, however, there have been very few books published on Entomology for beginners, but for those who wish to go more deeply into the subject than the present brief manual can pretend, we would strongly recommend them to obtain Mr. Kirby's excellent illustrated "Text-Book of Entomology" (Swan Sonnenschein & Co.), to which work we are greatly indebted for much valuable and interesting information.

# CONTENTS

# THE YOUNG COLLECTOR'S HANDBOOK

OF

# Ants, Bees, Dragon-Flies, Earwigs, Crickets, and Flies.

## INTRODUCTION.

By the Linnean system insects are divided into seven great Natural Orders—namely, Coleoptera, Orthoptera, Neuroptera, Hymenoptera, Lepidoptera, Hemiptera, and Diptera.

Of these, the Lepidoptera or Butterflies and Moths, and the Coleoptera or Beetles, have hitherto received the lion's share of attention, though this is less exclusively the case than it was some few years ago.

Any one who wishes to commence the study of these two orders now will have to work very hard indeed before he can hope to put anything new on record concerning them. There is, however, a very wide field for research and discovery open to all who will turn their attention to the other groups, and it is with the intention of encouraging the study of these " Neglected Orders " that we have written the present little handbook.

The orders which we here intend to treat of are the following :—

*The Hymenoptera*, including the Bees, Wasps, Ants, Saw Flies, Gall Flies, Ichneumon Flies, and their allies.

*The Neuroptera*, including the Dragon Flies, Day Flies, Lacewing Flies, Stone Flies, Caddis Flies, and their allies.

*The Orthoptera*, including the Grasshoppers, Locusts, Crickets, Cockroaches, Earwigs, and their allies.

*The Hemiptera*, including the Bugs, Skaters, Lantern Flies, Frog Hoppers, Aphides, and their allies.

*The Diptera*, including the Gnats, Midges, Crane Flies, Hawk flies, Bee Flies, Breeze Flies, Bot Flies, and their allies.

The number of insects is so large (about 13,000 different kinds are found in our own country alone) that it will be found impossible by any one to collect the whole at once; and to study them all is completely out of the question. A single insect, indeed, is ample

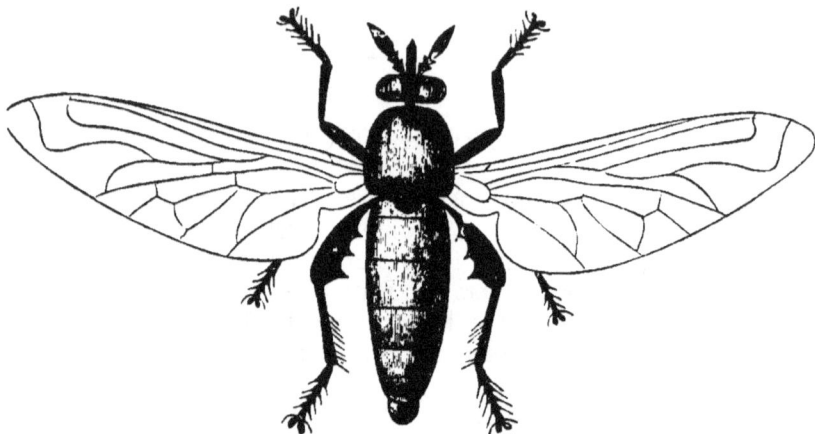

Fig. 1.—Hoplistomerus Serripes.

to occupy a whole lifetime in the elucidation of its life history. Most persons, when they commence to study insects, collect indiscriminately everything which comes in their way; but they soon find the subject too immense for them to grasp as a whole,

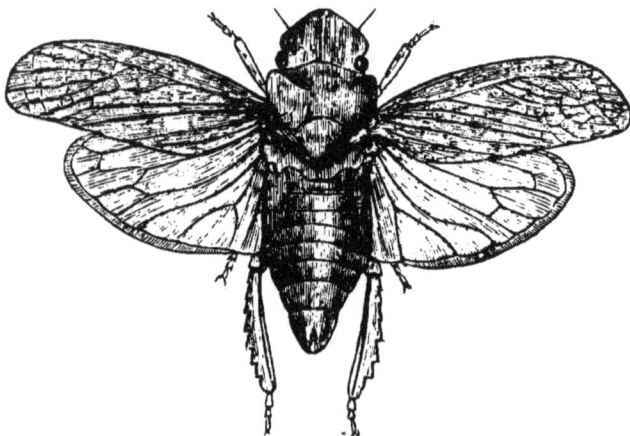

Fig. 2.—Ledra Aurita (Mag.)

and they either give up collecting altogether, or else confine their attention to a single group or order of insects. We would recommend every young person to make up his mind at the first which group or groups of insects he has a special fondness or liking for. Having done so, he will require to know the best means of attaining his object—namely, the formation of a collection. A few

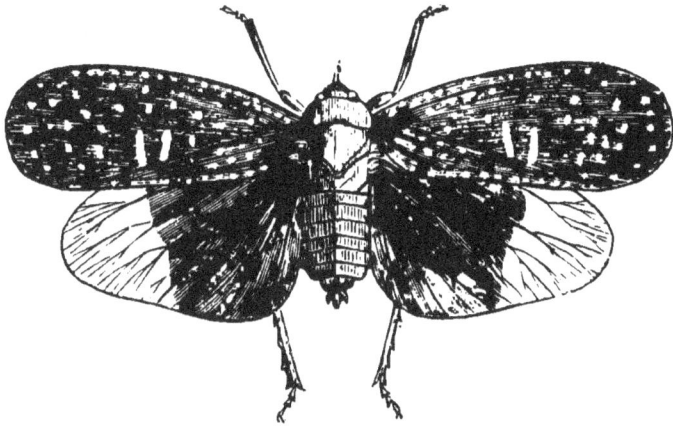

Fig. 3.—Aphæna Amabilis.

hints, therefore, as to the mode of collecting and preserving insects in general may be of assistance to the young collector.

First of all, as to the apparatus necessary for collecting insects.

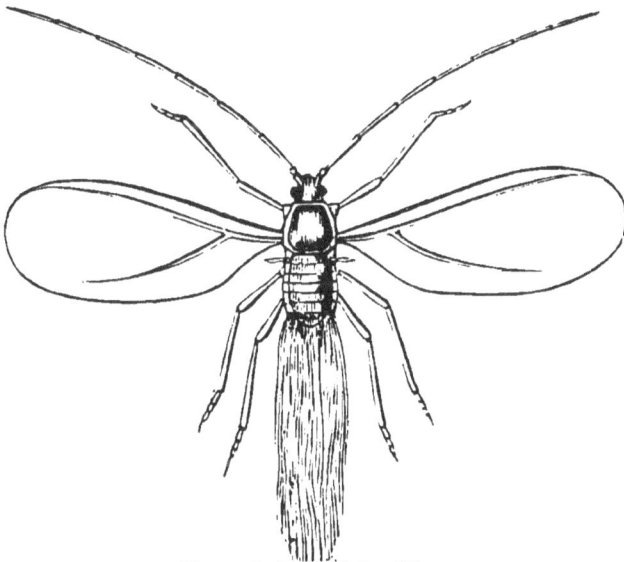

Fig 4.—Orthesia Urticæ (Mag.)

This will be found very simple and inexpensive. A butterfly net is, perhaps, the most useful requisite. An ordinary cane-gauze net will do well. A good stock of chip pill-boxes will also be needed.   These can be purchased from the shop of any Natural History dealer at about 3*d.* per dozen, nested (in four sizes).

For collecting aquatic insects and their *larvæ* we shall require a water net, made of fine zinc-gauze, about six inches in diameter, to group about among stones and other rubbish at the bottom of pools and ditches.   This instrument should be made so as to slip on and off a walking-stick when required.

A killing bottle charged with cyanide of potassium or with chopped laurel leaves completes the outfit.

Little need be said as regards the mode of collecting insects, as their habits may be gathered from the following pages.

They may be found almost everywhere, in woods, fields, gardens, in pools, ditches, canals, and rivers, under dead leaves and the bark of trees, among moss and stones, etc., etc.   They may also be found at all times of the year; even in the depth of winter some species may be met with.

Most insects may be preserved in the same manner as Butterflies and Moths.   They should be set on flat setting-boards, and left to dry thoroughly before being removed.

The smaller species may be gummed to cardboard in a similar way to Beetles.

In the arrangement of the insects in the cabinet the classification and order given in these pages may be followed.   About half-a-dozen of each species will be found a convenient number to collect.

# ORDER HYMENOPTERA.

The *Hymenoptera* belongs to one of the most interesting and extensive orders of insects. The structure and habits of the different species which it includes are very various. Their marvellous instincts have excited the admiration and wonderment of the philosophers of all ages. They are, indeed, by far the most intelligent of insects, being greatly in advance of any other group known to exist. The *Hymenoptera* are mandibulate insects, their mouths being formed for biting, and they undergo complete

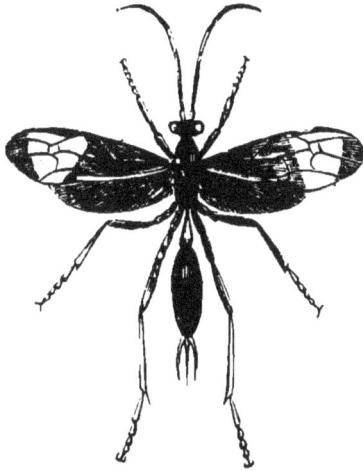

Fig. 5.—Cryptus Formosus.

metamorphoses. Perhaps the most striking external distinctive character is to be found in the structure of the wings, both pairs of which are membranous ; and another peculiarity of equal importance is the condition of the prothorax, which is reduced to very small dimensions. In the majority of the *Hymenoptera* the females are provided with stings, which serve many purposes, independently of weapons of defence.   Many families, however, do not possess stings.

Many of the *Hymenoptera* feed on plants and trees, but probably the greater number are parasitic on other insects.

This order includes the bees, wasps, ants, saw flies, gall flies, ichneumon flies, and many other familiar insects.

Fig. 6.—Pimpla Turionellæ (Mag.

The following will show the principal characteristics of each group into which the order has been divided.

> *Tribe* 1, *Aculeata.*—Generally social insects, consisting of males, females, and neuters.   Ovipositor modified into a sting.   *Larvæ*, footless grubs.
>
> *Tribe* 2, *Entomophaga.*—Mostly parasitic on the eggs and larvæ of other insects.   Ovipositor used as a borer. *Larvæ*, footless grubs.
>
> *Tribe* 3, *Phytophaga.*—Principally vegetable feeders.   Ovipositor used as a borer.   *Larvæ* having six or more legs.

### TABULAR VIEW
#### OF THE
### PRINCIPAL FAMILIES OF THE HYMENOPTERA.

#### TRIBE I. ACULEATA.

*Section* 1. *Anthophila.*

Family 1. Aphidæ or Honey Bees.
,,      2. Andrenidæ or Burrowing Bees.

*Section* 3. *Diploptera.*

Family 3. Vespidæ or Social Wasps.
,, 4. Eumenidæ or Bramble Wasps.
,, 5. Masaridæ or Solitary Wasps.

*Section* 3. *Fossores.*

Family 6. Philanthidæ or Bee-eating Wasps.
,, 7. Crabronidæ or Sand Wasps.
,. 8. Nyssonidæ or Fly-eating Wasps.
,, 9. Larridæ or Black Wasps.
,, 10. Sphegidæ or Grasshopper-eating Wasps.
,, 11. Pompilidæ or Burrowing Wasps.
,, 12. Bembecidæ or Scented Wasps.
,, 13. Sapygidæ or Bees' Nest Wasps.
,, 14. Scoliidæ or Beetle-eating Ants.
,, 15. Thymidæ or Stout-bodied Ants.
,, 16. Mutillidæ or Solitary Ants.

*Section* 4. *Heterogyna.*

Family 17. Formicidæ or Social Ants.

TRIBE II. ENTOMOPHAGA.

Family 18. Cympidæ or Gall Flies.
,, 19. Chalcididæ or Little Gall Flies.
,, 20. Proctotrypidæ or Bee Parasites.
,, 21. Braconidæ or Butterfly Parasites.
,, 22. Ichneumonidæ or Ichneumon Flies.
,, 23. Evaniidæ or Beetle Parasites.
,, 24. Chrysididæ or Golden Wasps.

TRIBE III. PHYTOPHAGA.

Family 25. Sericidæ or Tailed Wasps.
,, 26. Tenthredinidæ or Saw Flies.

TRIBE I.—ACULEATA.

The *Aculeata* contains the great majority of the *Hymenoptera.* Their principal characteristic is that the ovipositor of the female in most of the groups is modified into a sting. The *larvæ* are footless grubs. This tribe is divided into four sections, which are again subdivided into seventeen families, each of which it is our intention to discuss in their proper order.

*Section* 1.—*Anthophila.*

*Family* 1, *Aphidæ.*—This family contains the numerous species of honey bees which are familiar to all of us. They are very varied in their structure, colours, and habits.    Many species are

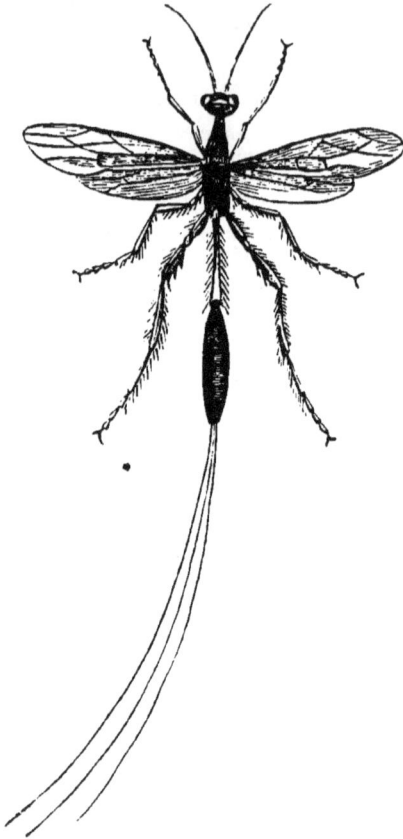

Fig. 7.—Megischus Annulator.

social, while many others are solitary.    The common Hive Bee
(*Apis mellifica*) belongs to the former class.    There is, perhaps,

Fig. 8.—Apis Mellifica (Queen).

no insect which has attracted so much notice as this species.
Volumes have been written respecting it, and philosophers in all

ages of the world's history have spent their whole lifetime in the elucidation of its economy.

In southern Europe, notably in Italy, a much brighter-coloured and finer insect is found, distinguished especially by having yellow

Fig. 9.—Apis Mellifica (Drone).

transverse bands on the abdomen. This bee was long supposed to be a distinct species, and was described under the name of *Apis ligustica*, but it is now regarded as merely a variety. This bee has been introduced into all the northern parts of Europe.

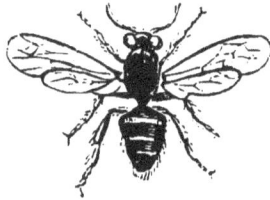

Fig. 10.—Apis Mellifica (Worker).

Several species of Humble Bees are very common in this country. One of the best known is the *Bombus terrestris*, the large females of which may attain a length of nearly an inch. This is a large black insect with the extremity yellow.

Fig. 11.—Melipona Anthidioides (Mag.)

In another rather smaller species, *Bombus lucorum*, the extremity of the abdomen is white. Both these species are sub-terranean bees, forming their nests in banks, etc.

2

Of the moss-builders, the best known, perhaps, is the *Bombus muscorum*, the largest specimens of which measure about two-thirds of an inch long.

Another species, *Bombus lapidarius*, is so called from a preference it shows for making its nests under stones. The end of the abdomen of this bee is bright orange-red.

Of the solitary bees a very common black species is *Anthophora*

Fig. 12.—Bombus Pratorum.

*acervorum*, which is usually found in abundance in the spring in the neighbourhood of banks and cliffs.

The violet Carpenter Bee (*Xylocopa violacea*), which chiefly inhabits the south of Europe, is a very pretty insect with violet-coloured wings.

The Mason Bee (*Chalicodoma muraria*) builds its nest, composed of fine grains of sand, very firmly united by a salivary

Fig. 13.—Xylocopa Violacea.

secretion, upon the surface of walls and similar situations. This species has hitherto not been met with in this country.

The Horned Bee (*Osmia bicornis*) is remarkable for the female having two little horns projecting from the front of her head. This insect usually burrows in sandy banks and cliffs. Another allied species (*Osmia hirta*) burrows in wood, whilst two others (*Osmia bicolor* and *aurulenta*) select ready-made nests in the shells of the common snails (*Helix hortensis* and *H. nemoralis*), within the whorls of which they build their cells of gnawed vegetable material.

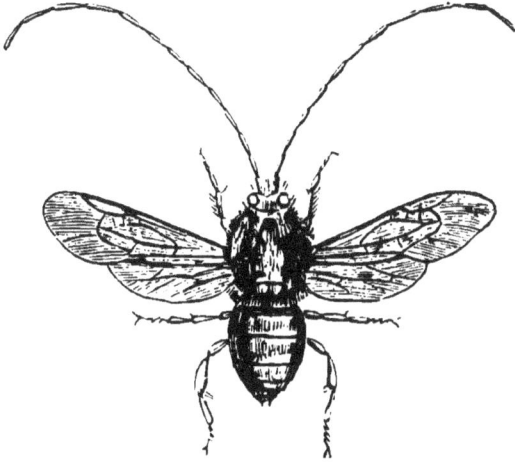

Fig. 14.—Ctenioschelus Latreillii.

The Leaf-cutting Bees, which belong to the genus *Megachile,* are also very interesting in their habits.

Fig. 15.—Csmia Tunensis.

*Family* 2, *Andrenidæ.*—All the insects belonging to this family are solitary in their habits. The species are very numerous in

Fig. 16.—Andrena Collaris.

this country. Many of them burrow in the ground, while others have been observed to make their nests in bramble stick. They are usually smooth, black insects, very unlike bees in appearance.

The females have no apparatus for carrying pollen either on the legs or on the abdomen.

### Section 2.—Diploptera.

*Family* 3, *Vespidæ.*—The Social Wasps, which belong to this family, are very well known everywhere in this country.   In their

Fig. 17.—Polistes Gallica.

general structure they resemble the bees, but are usually much more slender in appearance, and also much less hairy.

The general habits of these wasps are pretty uniform except in the matter of their architecture, and in this respect they display a remarkable variety.

Besides the common Wasp (*Vespa vulgaris*) two other species

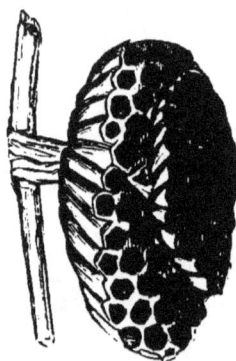

Fig. 18.—Nest of Polistes Gallica.

found in this country, which build their nests in the ground, follow the same principles in the construction of their nests.

The Hornet (*Vespa crabo*), which is remarkable for its large size, usually builds its nest in the hollow of a tree.   Both the hornet and common wasp sometimes build their nests under the eaves of houses or attached to a beam under the roof; and in these cases the outer covering of the nest is thinner and more delicate in texture than when the dwelling is exposed to the vicissitudes of the weather. Another common species inhabiting Britain is the Wood Wasp (*Vespa sylvestris*), which builds nests suspended from the branches of trees.

*Family 4, Eumenidæ.*—One of the commonest and best-known species belonging to this family is the Wall Wasp (*Odynerus*

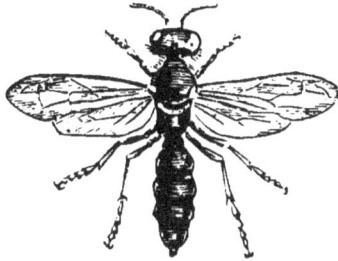

Fig. 19.—Odynerus Ovalis (Mag.)

*parietum*), which may be almost constantly seen haunting sunny walls during the months of June and July. It makes its burrows in walls and high banks, while many other allied species form their nests in the hollow stems of brambles.

*Family 5, Masaridæ.*—This family contains the great bulk of the solitary wasps, which, however, are principally inhabitants of warm climates.

The *Masaridæ* are a small group of black-yellow belted wasps, which are found in the south of Europe, but not in Britain, two of the commonest species being *Celonites apiformis* and *Ceramius fonscolombi*.

### Section 3.—*Fossores.*

*Family 6, Philanthidæ.*—Most of the species belonging to this

Fig. 20.—Cerceris Capito.

family are black with yellow spots and bands, and most of them are inhabitants of the warmer parts of the earth. Some of the species provision their nests with beetles and grasshoppers, while others attack bees, and are very mischievous, destroying great numbers.

*Family* 7, *Crabronidæ.*—This family includes a considerable number of solitary species of wasps. In colour they are generally black spotted and striped with yellow, but many of them are bright red.

The typical genus *Crabro* is a very extensive one, including over 150 species, a great proportion of which are inhabitants of Europe, while even Britain possesses more than thirty.

*Crabro cribrarius* is the largest British species. Its food consists principally of gnats and other dipterous insects. This insect and many others burrow in the ground, generally in hard sandbanks. Another species (*Crabro brevis*) frequents similar situations, and has been known to provision its nest with small beetles.

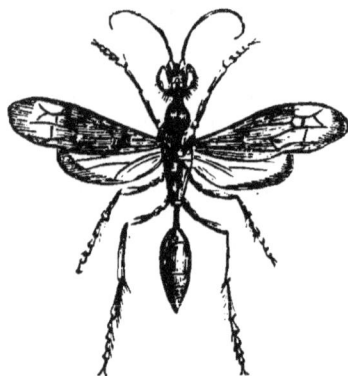

Fig. 21.—Chlorion Virid'æneum.

*Family* 8, *Nyssonidæ.*—This family is not a numerous one comparatively.

*Mellinus arvensis* is a black insect, about half an inch in length, with stripes on the abdomen and yellow legs. It provisions its nest with flies, and not being able to capture them by swiftness, runs past them when they are resting in an unconcerned manner till they are thrown off their guard, when they are pounced upon and carried off to the nest.

*Family* 9, *Larridæ.*—This family very much resembles the last. The species are mostly small insects, few of them measuring half an inch in length. They are usually black, sometimes with the abdomen red at the base.

*Family* 10, *Sphegidæ*—Most of the insects included in this family feed upon grasshoppers, and the manner in which they procure their prey is very curious. In attacking their bulky prey they use every endeavour to turn the grasshopper on its back. When they succeed in this they inflict stings in different parts of the underside of the abdomen and thorax, which soon paralyse the victim, which is then dragged to the nest of the ruthless destroyer.

*Sphex flavipennis* is a common species in the south of Europe.

*Family* 11, *Pompilidæ.*—Most of the species of *Pompilus* burrow in sand or sandy soil, and store their nests with spiders and the *larvæ* of insects.

One of our commonest species, *Pompilus fuscus*, is usually about half an inch long, and is black, with the first three segments of the abdomen red banded with black. This insect makes its appearance in the spring, and may be observed in sandy places throughout the summer. Another common British species is *Pompilus punctum*, which is black in colour.

*Family* 12, *Bombicidæ.*—This small family is not represented in

Fig. 22.—Sapyya Clavicornis.

Britain, though found in southern Europe. Many of the insects burrow in the sand, scratching a hole with their forefeet like a dog, as observed by Sir S. Saunders in the Ionian Islands. In the daytime they may be seen flying rapidly from flower to flower, and many of them exhale an odour of roses.

*Family* 13, *Sapygidæ.*—This is a small family containing only a single genus, with very few species.

The species of *Sapyga* occur in Europe and North America. They are supposed to be parasitic in the nests of bees, but the females of the common European species (*Sapyga pacca*) have been observed carrying small caterpillars, from which Mr. Smith (the great authority on Hymenoptera) justly infers that they are parasitic only to the extent of usurping the burrows made in sandbanks and dead wood by more industrious insects; their own structure not adapting them for the labour of digging.

*Family* 14, *Scoliidæ.*—Although this family is abundant in warm climates, we have only two small species belonging to the genus *Tiphia* in Britain.

They are black, with more or less reddish legs, and measure from a quarter to half an inch in length. In south Europe we meet with several large and handsome species, one of which, *Scolia hortorum*, is black with two yellow bands on the abdomen.

Fig. 23.—Scolia Hortorum.

*Family* 15, *Thymidæ.*—The insects of this family are almost exclusively confined to Australia and South America, where they are very numerous. They are generally of a black colour, with more or less extended yellow markings. They are very stout insects; in fact, their bloated bodies give them very little resemblance to any other insects, except perhaps to the Oil Beetles. Very little has hitherto been ascertained respecting their habits, but they are believed to be parasitic.

*Family* 16, *Mutillidæ.*—This family includes a large number of species, probably 1,500, but from the differences presented by

Fig. 24.—Mutilla Maura (Male).

the males and females, entomologists have found it difficult to arrive at any certainty upon this point. The species are spread over all the earth, but are particularly abundant in warm climates, where also, as usual, they attain the largest size and the most beautiful colouring.

One of the best-known species in our own country is *Mutilla europæa*, which is about half an inch long, of a black colour, hairy, with the thorax entirely red in the wingless females.

Fig. 25.—Mutilla Maura (Female).

This insect frequents the nests of Humble Bees, and its *larvæ* appear to be parasitic upon the *larvæ* of the bees.

## Section 4.—Heterogyna.

*Family* 17, *Formicidæ.*—To this family belongs the numerous species of ants, which are social insects, organised after the fashion of the bees and social wasps.

The number of species described is probably considerably over a thousand, but the total number must be very much greater if Mr. Bates is correct in his estimate that not less than 400 species inhabit the valley of the Amazon.

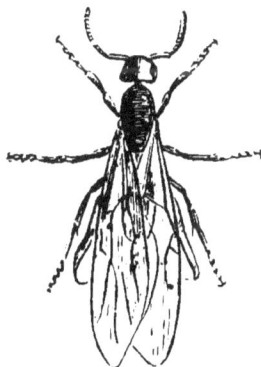

Fig. 26.—Formica Lignipeda (Male). (Mag.)

The habits of the ants are most interesting, and one of our greatest living naturalists, Sir John Lubbock, has devoted much of his time in elucidating their economy.

The nests are almost always chambered cavities, hollowed out either in the ground, in walls, and similar situations, or in dead and decaying wood.

One of the commonest examples in our own country is the Garden Ant (*Formica nigra*), which may be found everywhere in gardens making its nest in the ground.

Another common species is the pretty Turf Ant (*Formica flava*), which generally haunts commons and heaths, casting up small

Fig. 27.—Formica Lignipeda (Worker).  (Mag.)

hills, which serve to throw off the rain; and this species in some localities makes its nest under stones.   The Wood Ant (*Formica lignipeda*) is another familiar species.

A very large group of ants belong to the section *Myrmicinæ*, the best-known species of which are the Red Ants, *Myrmica rubra*, and their allies.

A very minute species which has been introduced into this country, probably from Brazil or the West Indies, is the Horse Ant (*Myrmica molesta*).   It is a very small brownish-yellow

Fig. 28.—Myrmica Rubra (Male).  (Mag.)

species, which seems to have been first observed in England in 1828.   It takes up its abode in houses, frequently in the neighbourhood of the kitchen fireplace, and when it multiplies becomes such a pest as to render the house uninhabitable.   Some of the metropolitan districts have been particularly infested.

## TRIBE II.—ENTOMOPHAGA.

Most of the insects belonging to this tribe are parasitic on other insects. The *larvæ* are footless. There are seven families included in the *Entomophaga*.

*Family* 18, *Cynipidæ.*—This family includes most of the gall flies. The number of species is very considerable. Of the great majority the females pierce with their ovipositor the tissues of plants and trees, and there deposit their eggs, from which the

Fig. 29.—Cynips Gallæ Tinctoriæ (Mag.)

*larvæ* are soon hatched. The irritation caused by this intrusion of a foreign body into the tissues produces the galls which are so commonly met with.

The galls produced by different species of flies differ greatly in form and structure. Some of them are round and smooth, like

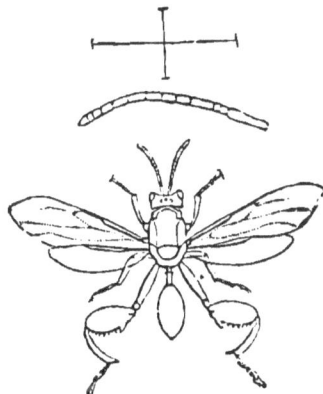

Fig. 30. –Smicra Sispes (Mag.)

fruits, such as the cherry gall of the oak leaves, produced by the puncture of *Cynips quercus-foli*.

The most singular, however, of all the galls is perhaps the Bedeguar, which is formed on the stems of wild roses by the puncture of a small species, *Rhodites rosæ*.

*Family* 19, *Chalcididæ.*—To this family belong many gall insects, principally found, however, in foreign countries.

The *Chalcididæ* include a great number of small species, few exceeding half an inch in expanse. Many of these are singular in shape, and others brilliantly metallic; but, owing to their small size, they have hitherto been studied by comparatively few entomologists.

Fig. 31.—Thoracamba Furcata (Mag.)

*Family* 20, *Proctotrypidæ.*—The *Proctotrypidæ* are probably much less numerous than the preceding family, but have been less studied, being generally smaller and more obscure in their habits; in fact, some of them share with several beetles the reputation of being the smallest insects.

Fig. 32.—Proctotrypes Rufipes (Mag.)

*Family* 21, *Braconidæ.*—This family is one of very great extent.

Many of the foreign species are rather large and handsome insects, often varied with black and yellow.

The best known of the *Braconidæ* is perhaps *Microgaster glomeratus,* a small blackish species with reddish-yellow legs, which destroys the *larvæ* of the common Cabbage Butterflies, round the dead body of which its little yellow cocoons may often be observed.

*Family 22, Ichneumonidæ.*—The *Ichneumonidæ* are rather large and slender insects, and are divided into many sub-families.

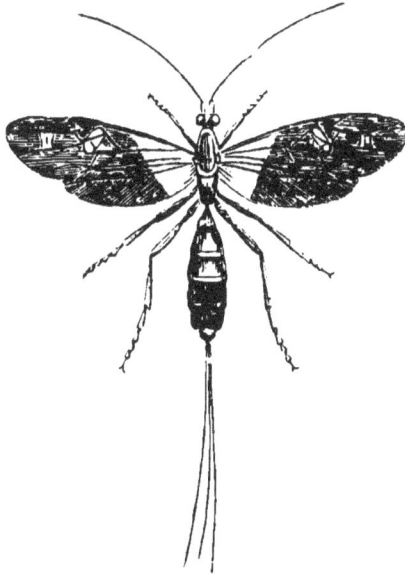

Fig. 33.—Bracon Bicolor.

They are a very numerous group. It has been calculated that nearly 5,000 species have been described, but the data generally are very untrustworthy.

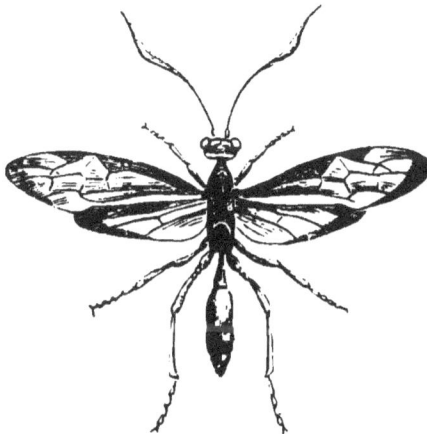

Fig. 34.—Joppa Antennata.

The species of *Trogus* are rather large insects, measuring an inch or more in length.   They are black with reddish legs and

Fig. 35.—Evania Appendigaster (Mag.)

abdomen, and the wings are sometimes slightly dusky at the
edges.

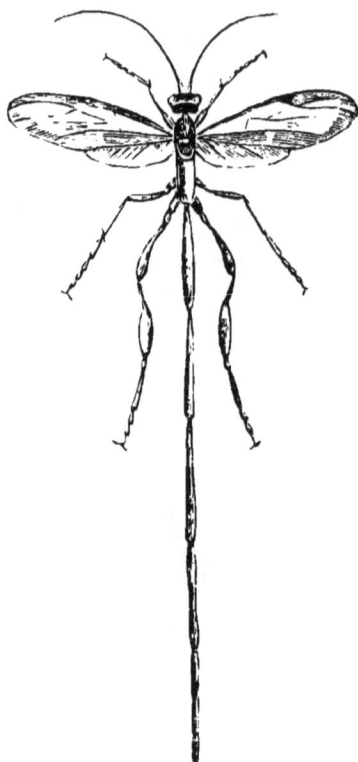

Fig. 36.—Pelecinus Politurator.

In another group, *Pimplinæ*, the ovipositor is generally very
long.   The best-known species is *Rhyssa persuasoria*, a blackish
insect, which measures about an inch in length.   This insect is

met with in fir plantations, and uses its extraordinary ovipositor to drill holes in trees infested by the *larvæ* of *Sirex gigas*, on which its own *larva* is parasitic. The insect frequently drives its ovipositor so firmly into the wood of the tree, that it is unable to withdraw it, and perishes in this position. ·

*Family* 23, *Evaniidæ.*—One of the most familiar insects belonging to this family is the *Evania appendigaster*. It is a small black insect, found in the south of Europe, and is parasitic on cockroaches. An allied British species, *Fœnus jaculator*, is a not uncommon insect found haunting the burrows of *Crabronidæ*, upon which it is probably parasitic.

*Family* 24, *Chrysididæ.*—The Ruby-tailed Wasps, or Golden Wasps, as the *Chrysididæ* are popularly called, are among the most brilliant of the *Hymenoptera*, most of the species being either of an intense green, blue, or fiery red.

They are small or moderate-sized insects, which are found on

Fig. 37.—Chrysis Ignita (Mag.)

walls or flowers in the full heat of the sun ; for, as a rule, the most brilliantly-coloured insects are diurnal in their habits.

As far as their habits are known, they deposit their eggs in the nests of other insects, chiefly *Hymenoptera*, on the *larvæ* of which their own offspring feed.

The commonest British species is *Chrysis ignita*, which is a very variable insect, both as regards size and colouring.

## Tribe III.—Phytophaga.

The insects belonging to the third tribe of the Hymenoptera are strictly vegetable feeders. There are only two families.

*Family* 25, *Siricidæ.*—This family includes the insects known as Tailed Wasps. It is not a very extensive family, and its species occur chiefly in Europe and North America, in both of which regions the typical genus *Sirex* is represented by large species.

The best-known European species which is common in some parts of Britain is the great Tailed Wasp (*Sirex gigas*), a very formidable-looking insect, of which the female often measures nearly an inch and a half in length.

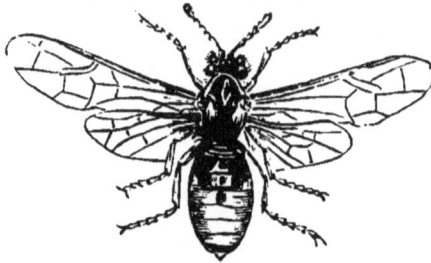

Fig. 38.—Cimbex Luteus.

The general tint is black with the antennæ, the sides of the thorax and the legs and apex of the abdomen yellow. This insect lives in pine and fir woods, and the female deposits her

Fig. 39.—Lophryus Pini (Mag.)

eggs in the woody parts of the trees, into which she bores to a depth of over half an inch by means of her long ovipositor.

Fig. 40. Pamphilius Faustus (Mag.)

Another species which occurs in this country is the *Sirex juvencus*, of a steel-blue colour, but smaller than the former.

*Family* 26, *Tenthredinidæ.*—This is a very extensive family and contains the numerous species of Saw Flies, so called because their ovipositor is in shape somewhat like a saw in appearance.

Probably the best-known species is the Gooseberry Saw Fly (*Nematus ventricosus*), whose speckled and green *larvæ* are so injurious in gardens and orchards. This insect is yellowish in colour, and about a quarter of an inch in length.

*Tenthredo æthiops*, a small black species, deposits its eggs upon fruit trees.

Many other species live on different kinds of plants and trees.

# ORDER NEUROPTERA.

INCLUDING THE DRAGON FLIES, LACEWING FLIES, DAY FLIES,
STONE FLIES, CADDIS FLIES, AND THEIR ALLIES.

---

Next to the butterflies and moths, the *Neuroptera* (or nerve-winged insects), to which belong the Dragon Flies, Lacewing Flies, Day Flies, Stone Flies. Caddis Flies, and their allies, are undoubtedly the most beautiful members of the insect tribe. Though they cannot compete with the *Lepidoptera* in point of colouring, it is questionable whether they are not more graceful and elegant in appearance.

The order *Neuroptera* was founded by the great Swedish

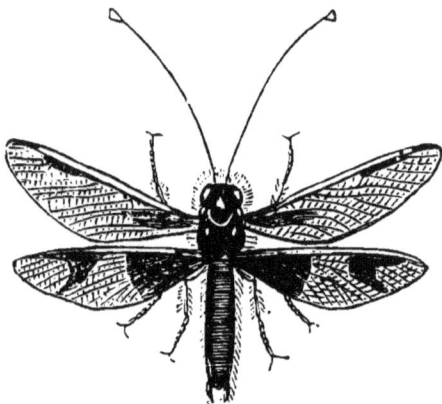

Fig. 41.—Ascacaphus Macaronius.

naturalist Linnæus, and its name has been kept intact since, though the classification of the insects comprised in it has undergone many modifications.

The *Neuroptera* are not a very numerous body of insects compared with either the *Coleoptera* or *Lepidoptera*, but they contain some of the largest and most handsome insects known. The order is very well represented in the temperate zones, though the finest species are met with in the tropical parts of the world. Nearly 700 species are known to inhabit the British Isles. In the whole world their number may be given in round figures at about 4,000, but there is no doubt that this number may be considerably increased.

The largest and most handsome species are the Dragon Flies or *Ornoptera*, one species occurring in our own country, measuring fully four inches in expanse of the wings. The Lacewing Flies (*Hemerobiidæ*), the Day Flies (*Ephemeridæ*), the Stone Flies (*Perlidæ*), and the Caddis Flies (*Trichoptera*), are among the other more conspicuous members of the group.

The *Neuroptera* may be divided into three great sub-divisions, according to certain well-defined characteristics. The modern

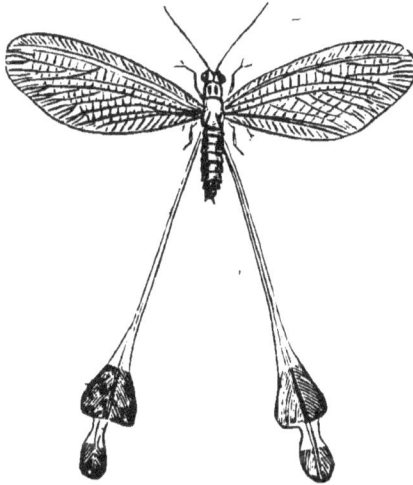

Fig. 42.—Nemoptera Extensa.

tendency is to raise each of these sub-divisions to the rank of orders.

The *Neuroptera* may be classified as follows :—

Sub Order 1, *Planipennia*, or true Neuroptera.

Metamorphoses complete, larvæ, mostly terrestrial.

Sub-Order 2, *Trichoptera*, or hairy-winged Neuroptera.

Metamorphoses complete, larvæ aquatic. Wings of Imago clothed with hairs.

Sub-Order 3, *Pseudo-Neuroptera*, or false Neuroptera.

Metamorphoses incomplete.

The most important distinction between the first two groups and the third is that the former undergo complete metamorphoses, whereas in the latter the transformations are incomplete or imperfect. The latter for this reason are often classified as a separate order.

The *Trichoptera* are also placed by some entomologists in a distinct order, on account of the peculiarity in their wings being clothed with hairs. They, however, present so many points both in habits and structure, analogous with the true *Neuroptera*, that we have retained them as a sub-division of the Order.

## TABULAR VIEW

### OF THE

## PRINCIPAL FAMILIES OF THE NEUROPTERA.

*Sub-Order 1.—Planipennia.*

Family 1. Myrmeleontidæ or Ant Lions.
,,    2. Hemerobiidæ or Lacewing Flies,
,,    3. Mantispidæ or Mantis Flies.
,,    4. Sialidæ or Sialis Flies.
,,    5. Raphiidæ or Snake Flies.
,,    6 Panorpidæ or Scorpion Flies,

*Sub-Order 2.—Trichoptera* (or Caddis Flies).

Family 1. Inæquipalpia or Large Caddis Flies.
Sub-family 1. Phryganeidæ.
,,    2. Limnophilidæ.
,,    3. Sericostomidæ.
Family 2. Æquipalpia or Little Caddis Flies.
Sub-family 1. Leptoceridæ.
,,    2. Hydropsychidæ.
,,    3. Rhyacophilidæ.
,,    4. Hydroptilidæ.

*Sub-Order 3.—Pseudo-Neuroptera.*

TRIBE I. ORNOPTERA (or Dragon Flies).

Family 1. Libellulinæ or Great Dragon Flies.
Sub-family 1. Libellulidæ or Libellulines.
,    2. Cordulidæ or Cordulines.
,,    3. Æschnidæ or Æschnines.
,,    4. Gomphidæ or Gomphines.
Family 2. Agroninæ or Slender Dragon Flies.
Sub-family 1. Agronidæ or Agronines.
,,    2. Calopterygidæ or Calopterygines.

TRIBE II.—SUBULICORNIA.

Family 3. Ephemeridæ or Day Flies.

TRIBE III.—PLECOPTERA.

Family 4. Perlidæ or Stone Flies.

TRIBE IV.—CORRODENTIA.

Family 5. Psocidæ or Book Mites.
   ,,   6. Embiidæ or Agile Mites.

TRIBE V.—SOCIALIA.

Family 7. Termitiæ or White Ants.

TRIBE VI.—PHYSOPODA (or Thrips).

Family 8. Tubulifera.
   ,,   9. Terebrantia.

TRIBE VII.—MALLOPHAGA (or Bird-lice).

Family 10. Philopteridæ.
   ,,   11. Liotheidæ.

TRIBE VIII.—THYSANURA (or Bristle-tails).

Family 12. Lepismidæ.
   ,,   13. Campodeidæ.
   ,,   14. Japygidæ.

TRIBE IX.—COLLEMBOLA (or Spring-tails).

Family 15. Smynthuridæ.
   ,,   16. Papyriidæ.
   ,,   17. Degeeriadæ.
   ,,   18. Poduridæ.
   ,,   19. Amouridæ.

SUB-ORDER 1.—PLANIPENNIA.

The *Planipennia* contains the most typical forms of the *Neuroptera*. This sub-order is sub-divided into six families, four only of which contain representatives in the British Isles. They are most

Fig. 43.—Myrmeleon Formicarius.

abundant in the tropical parts of the world, where also the most handsome and curious forms occur.

*Family 1, Myrmeleontidæ.*—The Ant Lions are the most familiar and important members of this family. The common Ant Lion

(*Myrmeleon europæus*), which is abundant in sandy places in the south of Europe, is a slender and elegant creature, with large finely reticulated rings. The *larva*, to which the name of "Ant Lion" properly belongs, is of a stout form and a greyish-yellow colour, covered with warty processes and with hairs. Its

Fig. 44.—Myrmeleon Larva.

food consists of ants and other small insects, which it captures by a singularly ingenious arrangement, namely, by means of a funnel-shaped pitfall which it constructs in the sand, and at the bottom of which it lies. When any unfortunate insect ventures too near, the Ant Lion sends up a shower of sand, and the victim in its

Fig. 45.—Nemoptera Coa.

consternation falls down the pit, where it is speedily seized and devoured.

Other species of ant lions are known to occur on the continent of Europe, but none hitherto have been discovered to inhabit this country.

*Family* 2, *Hemerobiidæ.*—The second sub-family contains the Lacewing Flies, many representatives of which occur in the British

Isles. They are among the most elegant and beautiful of the *Neuroptera*.

The Golden-eyed Fly (*Chrysopa vulgaris*) is a very abundant and well-known example, but notwithstanding its great beauty it is capable of emitting a very disagreeable odour when handled. This is a delicate green insect with a body half an inch long, and

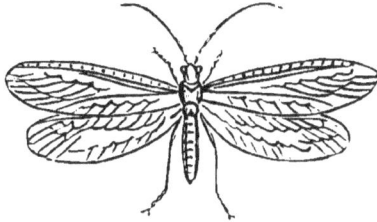

Fig. 46.—Chrysopa Septempunctata.

which may be seen almost everywhere on warm summer evenings, flying slowly about from tree to tree. The eggs, which are little round or oval bodies like small pearls, are deposited by the female in groups upon the leaves of plants and trees. The *larva* when fulfed is about half an inch in length. The food consists of plant lice and aphides, and it is thus a very beneficial creature, and should be encouraged by every gardener. The *pupa* of this insect is enclosed in a cocoon. Between thirty and forty species of lacewing flies are known to occur in our own country.

*Family* 3, *Mantispidæ.*—This family contains only one genus—

Fig. 47.—Mantis Pagana.

viz., *Mantispa*, or the Mantis Flies A single species, *Mantispa pagana*, is common in southern Europe, but does not occur in the British Isles.

*Family* 4, *Sialidæ.*—In the fourth family of the *Planipennia* the *larvæ* are aquatic in their habits with very few exceptions. The *pupa* is not enclosed in a cocoon as it is in the *Hemerobiidæ*. The single common British species (*Sialis lutaria*) is a blackish-brown insect, rather more than half an inch in length. It is well known to anglers, and may be found abundantly in the spring and early part of the summer upon walls and palings in the neighbourhood of water, and upon the stems and leaves of grasses and other plants growing in the water or upon its bank.

In repose the wings of this insect, as in the *Hemerolicidæ*, are laid over the back. They are sluggish creatures, and do not readily take to flight.

*Family* 5, *Raphiidæ.*—This family contains the curious group of insects called "Snake Flies" or "Camel Flies," which are

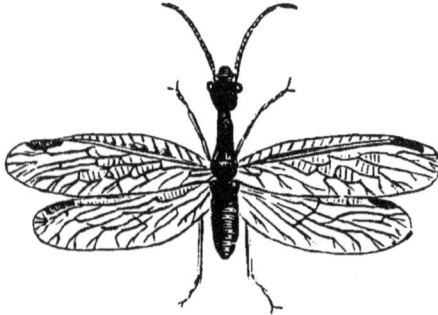

Fig. 48.— Raphidia Ophiopsis (Mag.)

included by some entomologists in the previous family. They have characteristics, however, which entitle them to be placed apart. They have a rather large head, which is attached to a greatly elongated prothorax by a thinnish neck, so that the head has considerable freedom of motion. The species are not

Fig. 49.—Panorpa Communis.

numerous, four kinds only occurring in Britain. The *larvæ* reside under the bark of trees, where they feed upon minute insects. The commonest species is probably *Raphidia mega-cephala.*

*Family* 6, *Panorpidæ.*—The *Panorpidæ*, or Scorpion Flies, are very curious creatures, characterised above all things by the perpendicularly placed and greatly elongated head.

The *larvæ*, so far as they are known, live in the earth, and are like caterpillars in their general form.

Five species occur in this country, the best known being

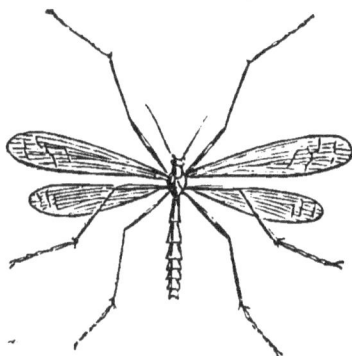

Fig. 50.—Bittacus Tipularius.

*Panorpa communis*, which may be met with almost everywhere about hedge banks and in lanes. It is about half an inch long. The wings are transparent, with dark brown spots, which are more or less confluent.

Another species, *Boreus hiemalis*, which possesses no wings, is found on the ground among leaves in the winter time. It does

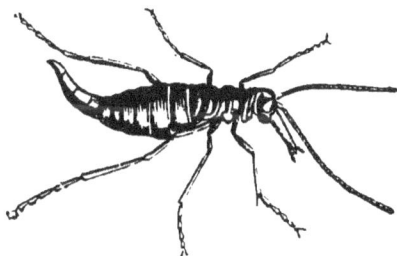

Fig. 51.—Boreus Hiemalis (Mag.)

not exceed one-sixth of an inch in length, and is of a metallic-green colour.

## Sub-Order 2.—Trichoptera.

The members of this group are the insects commonly known as Caddis Flies, and they are often ranked as a separate order by entomological writers.

Nearly 600 species of *Trichoptera* occur in Europe, of which about half the number are indigenous to the British Isles. The *larvæ* are aquatic, and when full-grown prepare for themselves curious dwellings composed of sticks, stones, and other materials wherein to pass the pupa state. They may be seen almost everywhere where there is water.

The caddis flies may be divided into two families, though the distinguishing marks are so minute that they really ought to be

Fig. 52.—Hydropsyche Montana.

classified as one family only. To Mr. McLachlan we owe many thanks for his researches in this group of insects.

*Family* 1, *Inæguipalpia.*—This family includes the largest species of the tribe, and are principally found in northern regions.

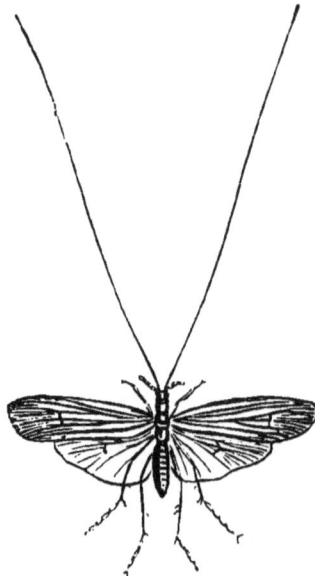

Fig. 53.—Marronema Rubiginosa.

One of our largest British species is *Phryganea grandis*, which measures four-fifths of an inch in length, and over two inches in expanse of the wing. It is an abundant insect, of a brown colour, with yellow rings on its *antennæ*, and the anterior wings are ash-coloured, clothed with brown.

The *larvæ* of this group for the most part inhabit quiet waters, ponds, canals, etc.

This family is subdivided into three sub-families, viz., *Phryganeidæ, Limnophilidæ, Sericostomidæ.*

*Family* 2, *Æguipalpia.*—This family constitutes a second group, and also contains many species, many of them, however, being very minute insects, some hardly one-eighth of an inch across the wings. They make little cases of silk resembling seeds, to the outer surface of which they attach grains of sand, etc.

The *Æguipalpia* contains four sub-families—viz., *Leptoceridæ, Hydropsychidæ, Rhyacophilidæ,* and *Hydroptilidæ.*

Of the *Leptoceridæ, Molanna augustata* may be taken as a typical example, the *larva* of which lives on the sandy bottom of pools, and is very difficult to detect.

## Sub-Order 3.—Pseudoneuroptera.

In the third sub-order of the *Neuroptera* are grouped together a series of insects which present great divergences of character, and really do not belong to the true *Neuroptera* on account of their incomplete metamorphoses They, however, for the most part, resemble the *Neuroptera* in the structure of their wings.

They are divided into several tribes and many families.

## Tribe I.—Ornoptera or Dragon Flies.

To this tribe belong the Dragon Flies, the largest and most beautiful members of the whole order.

About 1,500 species have been described from various parts of the world, and of these about fifty are known to inhabit our own country.

Their habits are very much alike. The insect passes all the earlier stages of its existence in water. The *larvæ* are most voracious creatures, and are undoubtedly the most predaceous of insects. The apparatus by which they capture their prey consists of a peculiar modification of the *labium.*

When full grown the *larvæ* crawl up the stem of some aquatic plant out of the water, and after resting there for a longer or shorter time the skin splits open along the thoracic region, and the perfect insect by degrees struggles out of its investment, and when the wings are dried it starts off to continue the same scene of rapine which has characterised its subterranean existence.

The perfect insect may be seen hawking about for insects in the neighbourhood of pools in all fine weather during the summer

and autumn months. In dull weather, however, they usually remain
at rest on the leaves of plants and trees, etc. The eyes of Dragon
Flies are most beautiful objects when viewed under the microscope;
they are composed of a great number of facets or lenses. In
one species of Dragon Fly as many as 10,000 of these facets have
been counted in each of its eyes.

Besides these compound eyes most dragon flies have additional
eyes, called *Ocelli*, which are situated on the top of the head;
they are, however, quite simple.

It is commonly thought by persons who are not naturalists that
dragon flies sting; such an erroneous idea we take the opportunity
to correct.

The *Ornoptera* are divided into two principal families.

Fig. 54.—Libellula Depressa.

*Family* 1, *Libellulinæ.*—To this family belong all the great
dragon flies, mostly thick-bodied insects. It is sub-divided into
four sub-families.

To the first sub-family, the *Libellulidæ*, many common and
familiar dragon flies belong. The best known is what is vulgarly
called the " Horse-stinger " (*Libellula depressa*), an insect about
two inches long, with a rather depressed abdomen, which is
yellowish-brown, with yellow spots on the sides in the female, and
coated with a beautiful violet-blue powder in the male.

It may be seen almost everywhere, hawking for flies about
rivers and ponds, during warm weather.

The second sub-family, the *Cordulidæ*, contains four British
species, of which the beautiful *Cordulia metallica* is the typical
example.

The *Æschnidæ* contain some of the largest members found. The great dragon fly (*Æschna grandis*) is one of these. It is nearly three inches long and four inches in expanse of the wings, and is of a light rusty-brown colour with a few pale markings.

Of the *Gomphidæ* only four species inhabit our own country.

*Gomphus vulgatissimus* is a black insect, nearly two inches long, with yellow bands on the thorax, and a line of the same colour along the back of the abdomen.

*Family* 2, *Agroninæ.*—This family contains a number of slender-bodied insects. There are two sub-families.

The *Agrionidæ* are a very numerous group. The typical form,

Fig. 55.—Calopteryx Virgo (Male).

*Agrion puella*, which is a beautiful, slender creature, measures about two or two and a half inches in expanse of the wings. The abdomen of the male is banded with azure blue, that of the female being brassy-black.

The *Calopterygidæ* contain only two British species—viz., *Calopteryx virgo* and *C. splendens*.

### TRIBE II.—SUBULICORNIA.

The second tribe of the *Pseudo-neuroptera* contains a group of insects which are so similar to one another in form that they are all included in one family.

*Family* 3, *Ephemeridæ.*—The *Ephemeridæ*, or Day Flies, as they are popularly called, are delicate, elongated, soft-bodied creatures, with a moderately small head. The antennæ, which spring from the forehead below the ocelli, are short and awl-shaped. These insects, which seem to be found mostly in temperate climates, are remarkable for the great delicacy of their structure, and for the extreme shortness of their lives in the perfect state, which seems in general scarcely to exceed a day.

Nearly fifty species of *Ephemeridæ* occur in this country. The best known is *Ephemera vulgata*, of which the sub-imago is called the " green drake," and the imago the " grey drake " by anglers. The two-winged *Clöen diptera*, which, however, is a much smaller species than the foregoing, is also very abundant.

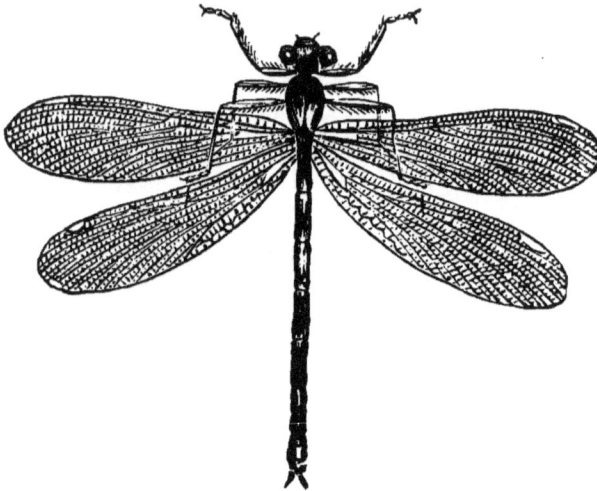

Fig. 56.—Calopteryx Virgo (Female)

TRIBE III.—PLECOPTERA.

The *Plecoptera* contain only one family.

*Family 4, Perlidæ.*—The *Perlidæ*, or Stone Flies, are chiefly inhabitants of the temperate regions ; they are of small size, and not very numerous in species, about thirty being indigenous to the British Isles.

The *larvæ*, which are aquatic in their habits, closely resemble the perfect insect in form. Their food consists of other insects and they prey especially upon the *larvæ* of the Day Flies.

The perfect insects are usually found resting quietly on plants and other objects on the banks of streams. The large brown species, *Perla bicaudata*, appears in the spring, and is very common in the neighbourhood of pools, streams, and canals.

### TRIBE IV.—CORRODENTIA.

This tribe contains two families, one of which is represented in the British Isles.

*Family* 5, *Psocidæ.*—This is a family of small insects with simple veined wings. They have a rather large head, and bear a pair of long tapering antennæ. They are found upon trunks of trees, old palings, walls, etc. They are very active in their

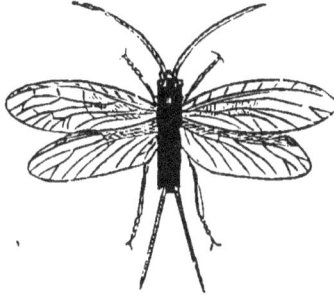

Fig. 57.—Perla Bicaudata.

movements, and generally appear in the imago state about the end of summer or beginning of autumn.

*Atropos pulsatorius* is one of these. It lives in books and among old damp papers, whence it is often called the "Book Louse."

Fig. 58.—Psocus Bipunctatus.

They are also very destructive to collections of preserved insects and plants. About thirty species occur in this country.

*Family* 6, *Embiidæ.*—This family contains no representative in the British Isles, and consists of a small number of almost exclusively exotic species. Two species, however, inhabit the continent of Europe, one of which, *Embia Solieri*, occurs in the south of France on the Mediterranean littoral.

These insects are very closely allied to the *Psocidæ* in their habits.

## Tribe V.—Socialia.

The Socialia contains a very peculiar group of insects closely allied to the cockroaches in the *Orthoptera*. They may be considered a connecting link between that order and the Neuroptera. There is only one family.

*Family* 7, *Termitiæ.*—The *Termitiæ*, or White Ants, are almost

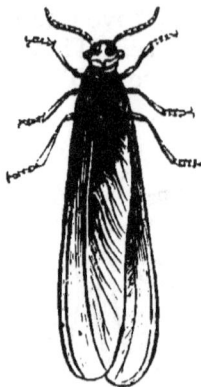

Fig. 59.—Termes Bellicosus (Male).

exclusively inhabitants of tropical climates, and are unrepresented in our own country.

They are a very remarkable family of insects. They build most complicated dwellings, consisting of innumerable galleries

Fig. 60.—Termes Bellicosus (Soldier).

and chambers, and they are so interesting in their habits that every traveller who has seen anything of them has always a great deal to relate concerning them.

One species, *Termes lucifugus*, inhabits the south of France, and is very well known on account of the mischief which it sometimes causes.

## Tribe VI.—Physopoda, or Thrips.

This tribe of insects, to which the name *Thysanopiera* is sometimes applied, contains many curious forms.

There are two principal groups of these insects—namely, the *Tuberlifera* and the *Terebrantia*. They are generally known to gardeners by the name of "Thrips." Some of them occasionally prove injurious to cultivated plants. This is especially the case with the Corn Thrips (*Thrips cerealium*), which generally attacks the ears of corn, and, when numerous, is very mischievous.

### TRIBE VII.—MALLOPHAGA OR BIRD LICE.

These insects, which are known as "Bird Lice," were formerly placed among the true lice, but they differ in the possession of biting mouths, and in the diet to which such a structure adapts them.

A great number of these curious little insects have been

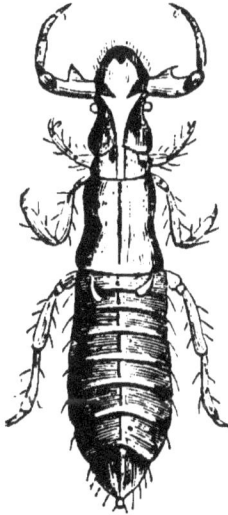

Fig. 61.—Lipeurus Diomedeæ (Mag.

recorded, and they inhabit all parts of the world. They live among the feathers of birds and the hairs of mammalia.

Almost every animal and bird is subject to these parasites.

The common fowl, duck, goose, game birds of all kinds, and pigeons, are very commonly infested by them, as are also the dog, the cat, the sheep, and the guinea-pig.

The two principal families of the *Mallophaga* are the *Philopteridæ* and the *Liotheidæ*.

### TRIBE VIII.—THYSANURA OR BRISTLE TAILS.

The forms composed in this tribe of insects are reckoned at present to be the nearest resemblance to the theoretical progenitors

4

of the insects; in fact, Sir John Lubbock hints that they might well be regarded not as insects at all, but rather as the surviving

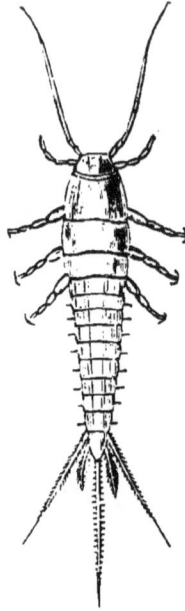

Fig 62.—Lepisma Saccharina (Mag.

representatives of a group formed by the ancestors of the whole multitude of insect types.

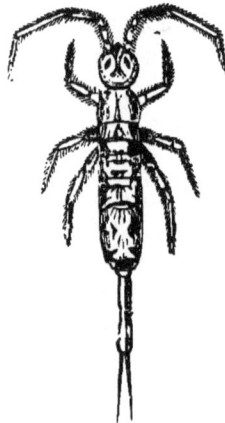

Fig. 63.—Orchesella Rufescens (Mag.)

The food of these creatures consists of decayed vegetable matter.

Three principal families may be recognised—viz., *Lepesmidæ*, *Campodeidæ*, and *Japygidæ*.

## TRIBE IX.—COLLEMBOLA OR SPRING TAILS.

These insects closely resemble the previous tribe in general character, to which also the same remarks nearly apply.

They are generally small insects, a length of a quarter of an inch being considerably above the average. They are found commonly in loose earth, under decaying leaves in woods, in moss,

Fig. 64.—Isotoma Gervaisi (Mag.)

under the bark of dead trees, and in rotten stumps. They always prefer damp situations.

Cold seems to have but little effect upon them ; they will recover their activity after being frozen. One species, *Desoria glacialis*, is found enjoying itself upon the Swiss glaciers, and another, *Degeeria nivalis*, occurs upon the surface of snow in many parts of Europe. Some species also may be met with hopping about upon the surface of standing water. *Podura aquatica*, a minute blue-black species, is common in such situations in England.

The five principal families are the *Smynthuridæ*, *Papyriidæ*, *Degeeriadæ*, *Poduridæ*, and the *Amouridæ*.

# ORDER ORTHOPTERA.

INCLUDING THE GRASSHOPPERS, LOCUSTS, CRICKETS, COCKROACHES,
EARWIGS, AND THEIR ALLIES.

----

The *Orthoptera* include all those forms of insects which have
four wings, the anterior pair being of a leathery nature. They
are mandibulate ; that is, their mouths are formed for biting, and
they undergo an imperfect metamorphosis.

The insects comprised in this order are extremely numerous
and destructive in the tropical parts of the world (Kirby gives
their number as about 7,000), but they are only represented in
the British Isles by about sixty species, few of which are really
abundant. They include the cockroaches, crickets, grasshoppers,
locusts, earwigs, etc.

All these insects may be divided into three sub-divisions or
tribes—namely, the Runners or *Cursoria*, the Leapers or *Saltatoria*,
and the Earwigs or *Euplexoptera*.

The following table will show at a glance the outlines of the
proposed arrangement :—

*Tribe* I., *Cursoria.*—Hind wings with reins radiating from the
base. Hind legs formed for running.

*Tribe* II., *Saltatoria—*Hind wings with veins radiating from
base. Hind legs formed for leaping.

*Tribe* III., *Euplexoptera.*—Hind wings with veins radiating
from the apex of a horny piece occupying the base of
the anterior margin.

The *Euplexoptera*, or Earwigs, are regarded by some authors as
a separate order. They are thus treated by Westwood in his
" Introduction to the Modern Classification of Insects," but the

modern tendency is to revert to the Linnæan system, and retain them among the *Orthoptera.*

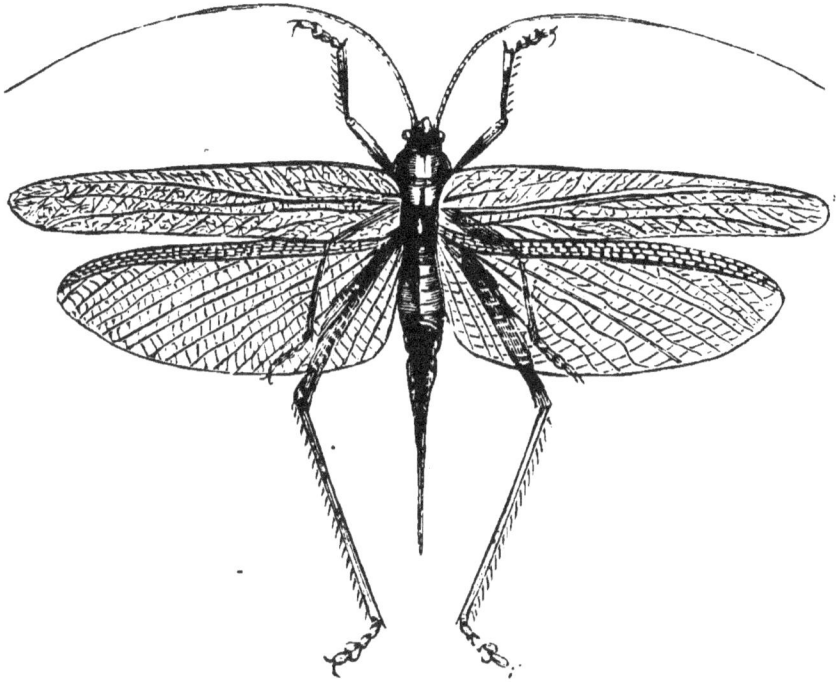

Fig. 65.—Phasgonura Viridissima.

## TABULAR VIEW

OF THE

## PRINCIPAL FAMILIES OF THE ORTHOPTERA.

TRIBE I.—CURSORIA.

Family 1. Blattidæ or Cockroaches.
　,, 　2. Mantidæ or Praying Insects.
　,, 　3. Phasmidæ or Stick and Leaf Insects.

TRIBE II.—SALTATORIA.

Family 4. Achetidæ or Crickets.
　,, 　5. Gryllidæ or False Locusts.
　,, 　6. Locustidæ or True Locusts and Grasshoppers.

TRIBE III.—EUPLEXOPTERA.

Family 7. Forficulidæ or Earwigs.

We will now proceed to describe each of these families in turn.

## TRIBE I.—CURSORIA.

The insects constituting this group are distinguished by having their hind legs adapted for walking or running. They are subdivided into three families.

*Family* 1, *Blattidæ.*—This family includes the numerous species of cockroaches, or "black bats" as they are often called by uneducated persons. They are represented in all parts of the world, but are most abundant within the tropics, and especially in central and southern America, where also the largest and finest species are to be found. Many curious forms are also to be met with in India and different parts of Africa.

They are very active creatures, and run with considerable rapidity, but their activity is chiefly nocturnal, and during the day they generally remain quietly concealed in some obscure retreat.

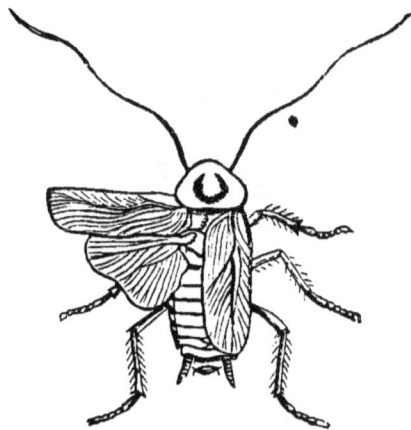

Fig. 66.—Blatta Orientalis.

Their diet consists of both vegetable and animal matter, and offal of every description. They are most useful in hot climates, acting the part of scavengers, and by these means preventing the outbreak of malaria and other pestilential diseases.

The best-known species in this country is the common Cockroach (*Blatta orientalis*), which, by-the-bye, is not a native of Europe, having been introduced from the East several centuries ago, and to have made its home here.

Other species besides this are met with occasionally in different parts of the country, having been introduced in like manner at different times from foreign parts. The Giant Cockroach (*Blaberus giganteus*) is one of these. Its native home is South America and the West Indies, and in the latter place it is commonly called the "drummer." It measures nearly three inches in length.

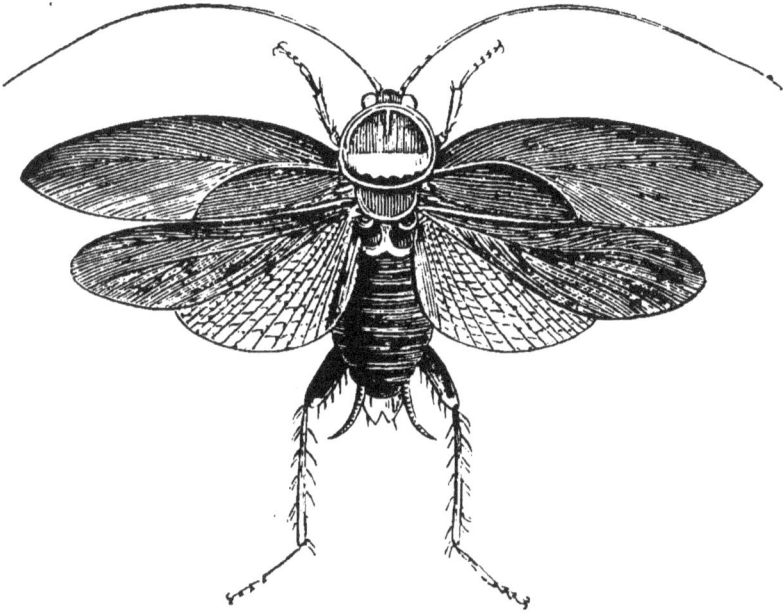

Fig. 67.—Blaberus giganteus.

The Zoological Gardens in London is a favourite resort of this species.

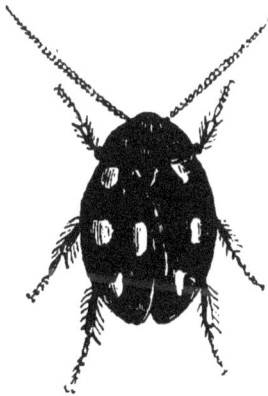

Fig 8.—Corydia Petiveriana.

*Family 2, Mantidæ.*—The *Mantidæ* may be at once distinguished from the insects comprising the other two families of the *Cursoria*

by the structure of the forelegs, which are converted into powerful
raptoral organs.

The body of these insects is more or less elongated, and the
head, which is triangular or heart-shaped, is attached to the
thorax by a distinct neck.

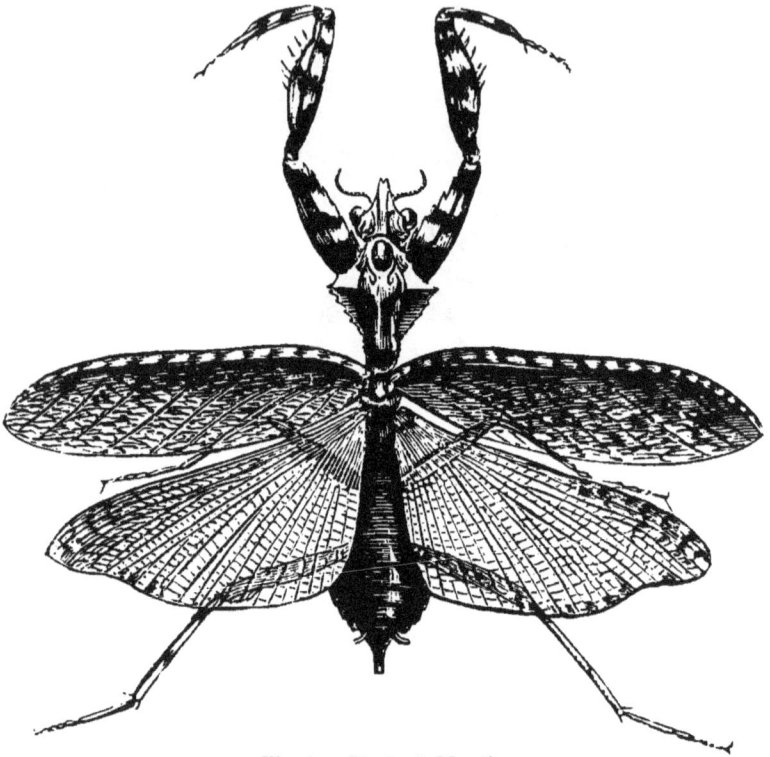

Fig. 69.—Blepharis Mendica.

The *Mantidæ*, or Praying Insects, are celebrated for their habit
of resting on their four hind legs, and with their front legs raised
in the air, in what was formerly supposed to be an attitude of
devotion, but really in an observant attitude, and on the alert for
their prey.

None of these insects are met with in the British Isles, but
several species are rather abundant in the south of France.   Two
species, *Mantis pauperata* and *Mantis religosa*, measure about two
or two and a half inches in length.   A third species, *Mantis
oratoria*, is also common but somewhat smaller in size.

*Family* 3, *Phasmidæ.*—These insects very much resemble the
*Mantidæ* in general appearance, to which also they are very closely
allied.

They are among the most peculiar insects in existence. Their appearance is comical in the extreme. Many of them resemble sticks, either green growing trees or brown and withered branches, and hence the name of stick insects commonly applied to them is very appropriate. On account of their skeleton-like forms they have also been likened to ghosts and spectres.

Some few species mimic leaves of trees and various plants, and these are often called " leaf insects."

The number of species of this family is not very large, and by far the greater part of these are inhabitants of the warmer regions of the earth, and they seem to increase in size, especially the nearer their home lies to the Equator.

Among the more interesting species we may mention *Cyphocrania semirubra* from Brazil, with short greenish elytra and pink wings.

*Bacillus rosii* is a brown, wingless form found in Italy and the south of France.

Some of the tropical species are among the largest insects known. A winged Australian species attains the length of ten inches.

*Lopaphus cocophagus* is a common species in the South Sea Islands, and sometimes commits great ravages in the plantations of cocoa-nut trees. When this insect is alarmed it squirts out a highly acrid fluid, which causes great pain, and sometimes blindness if it reaches the eye.

We may remark that no species of *Phasmidæ* are known to inhabit the British Isles.

## TRIBE II.—SALTATORIA.

The principal character of the insects belonging to this tribe consists in the adaptation of the hind legs to the purpose of leaping.

The males of most of the species possess the faculty of producing loud chirping sounds, but the means by which this is effected vary in the different families.

*Family* 4, *Achetidæ.*—The most familiar British insect belonging to this family is the common House Cricket (*Acheta domestica*). Living, as it does, in the immediate vicinity of the fire, it seems to be totally independent of the changes of the seasons, and may usually be found of all ages at all periods of the year. Crickets are particularly abundant in bakehouses, being extremely partial to warmth. The chirping noise which they produce at night-time is sometimes almost deafening, and it is really surprising from what a very long distance the sound may be heard.

Their food consists of bread crumbs and scraps of meat and vegetables which have been allowed to accumulate during the daytime  The best way, therefore, to get rid of the insects is to

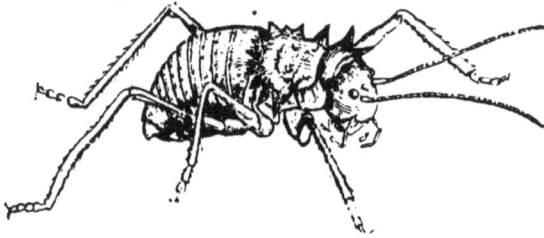

Fig. 70.  Hetrodes Pupa.

take proper care in clearing away all the refuse after meals, for if this were not there the crickets could not subsist, and would soon quit the house.  They are, however, generally objected to only by

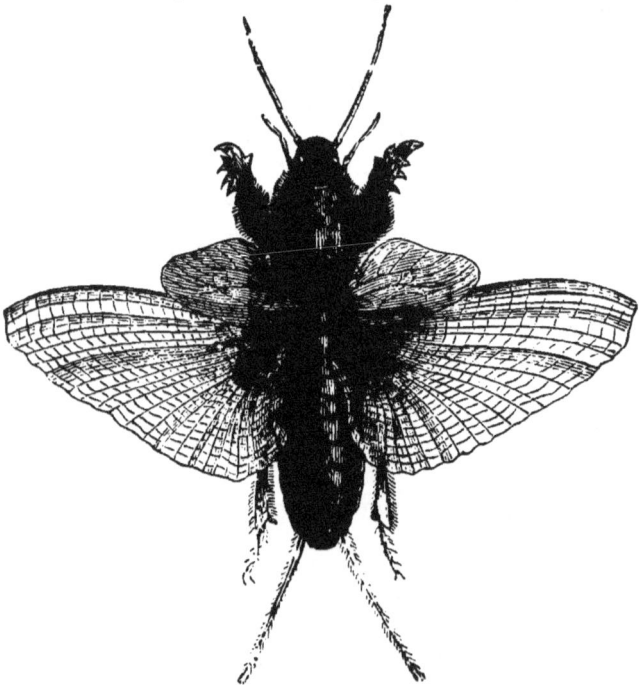

Fig. 71.—Gryllotalpa Vulgaris.

weak-nerved people, and do very little harm, if any whatever.  It is a question whether they are not very useful in acting as scavengers.

A nearly allied British species is the Field Cricket (*Acheta cam-*

*pestris*), which is rather larger than the house cricket. It is comparatively rare and local in this country, but abounds everywhere on the Continent.

A third British species is the Wood Cricket (*Nemolicus sylvestris*), which is considerably rarer than either of the preceding. It is found abundantly among dead leaves in woods in France and other parts of the Continent, but is very rarely found in England.

The Mole Cricket (*Gryllotalpa vulgaris*) is perhaps the most interesting member of the family. It is a large, robust insect, about an inch and a half in length and of a very dark brown colour. It is remarkable for the peculiar shape of its front legs, which exactly resemble those of the mole.

It burrows in loose soil, and, like the mole, it passes along close beneath the surface of the ground, and often raises a small ridge

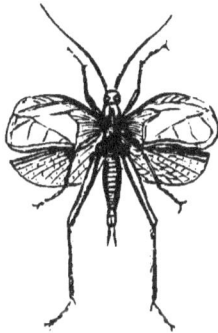

Fig. 72.—Œcanthus Pellucens.

as it advances. It frequents gardens, especially near the banks of canals, and is also fond of damp meadows and other localities in the vicinity of water.

The eggs to the number of 200 or 300, are deposited in a chamber of considerable dimensions, and enclosed in a sort of cocoon-like envelope.

The *larvæ* when first hatched are white, and are said to be three years in arriving at a state of maturity. The mole cricket is found chiefly in the south of England.

*Family* 5, *Gryllidæ*.—The most conspicuous insect belonging to this family in Britain is known as the great Green Grasshopper (*Phasgonura viridissima*), which measures nearly four inches in expanse of the wing, and is therefore nearly as large as the migratory locust which sometimes visits us.

The great green grasshopper is not, however, a very common

insect in this country, being principally confined to the south of the island.

Another European and British species, *Decticus verrucivorus*, is of about the same length. It receives its name from the custom prevailing among the Swedish peasants of making it bite their warts. This insect, in common with many others of the same family, when at all roughly handled, emits from the mouth a

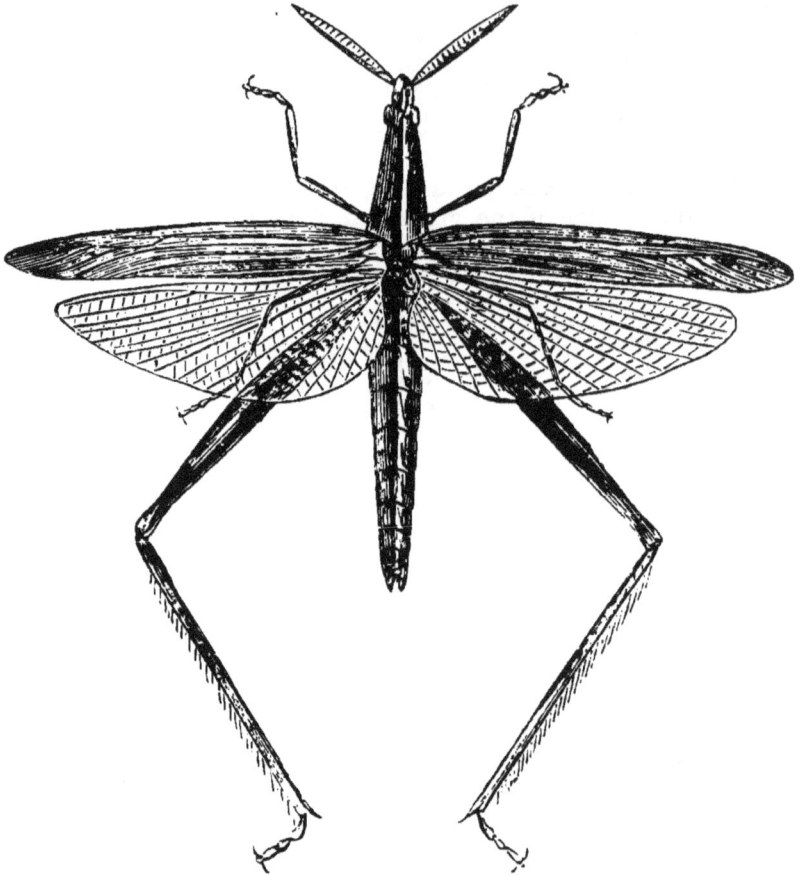

Fig. 73.—Truxalis Nasuta.

brownish fluid, which is said to possess acrid qualities, and the introduction of this into the warts is supposed to cause their disappearance.

*Family* 6, *Locustidæ.*—This family, which includes the grasshoppers and true locusts, is easily distinguished from both the preceding families by the character of the antennæ, which are comparatively short.

The common English Grasshopper (*Rhammatocerus biguttulus*), whose song must be familiar to every one who has walked through the fields during the summer time, is produced by the friction of the hinder thighs against the wing-cases.

The locusts have been celebrated from the dawn of history on account of the terrible ravages which they have committed in

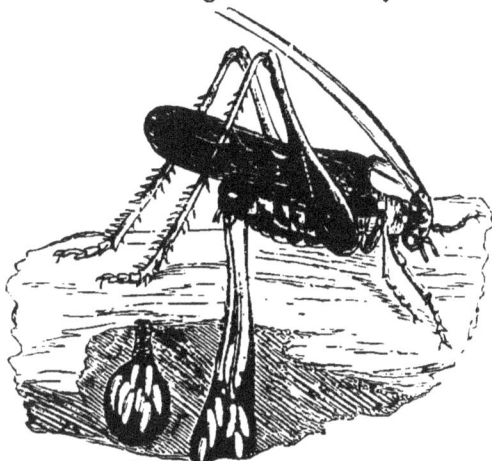

Fig. 74.—Decticus Verruciverus.

various parts of the world; and although many different species have made their appearance in our own country at various times, the climate, fortunately, seems unsuitable for them to live in, and they have, therefore, not been known to breed here.

The Migratory Locust (*Ædipoda migratoria*) is perhaps the best

Fig. 75.—Locusta Peregrina.

known in this country on account of its visits, many stray individuals of which have visited us at different periods.

Locusts are known all the world over, in fact, nearly every country has a species peculiar to itself. The most extraordinary accounts are on record of the vastness of the swarms which every now and again invade particular districts. They are said sometimes to absolutely darken the sun.

They clear off everything from the surface of the ground as completely as if the place had been visited by fire.

In many eastern countries locusts are relished as articles of food.

The distinctive character of the insects comprised in this group is found in the structure of the wings. The anterior pair are of a horny or leathery consistence, but always much shorter than the

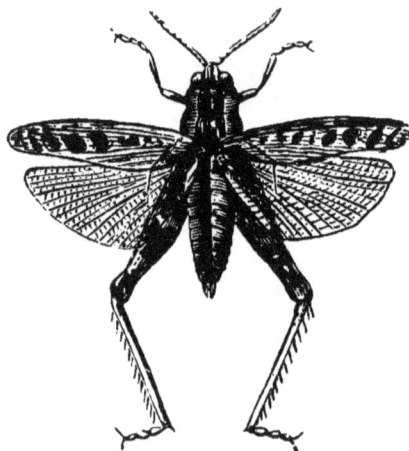

Fig. 76.—Caloptenus Italicus.

abdomen. The hind wings, on the contrary, are of large size, and are composed of a very delicate membrane. There is only one family.

### TRIBE III.—EUPLEXOPTERA.

*Family* 7, *Forficulidæ.*—The insects of this family present a great uniformity of structure. They are, for the most part, of nocturnal habits, concealing themselves during the day in crevices under the bark of trees, or in the ground under stones and leaves, etc.

Their food consists almost exclusively of vegetable substances, and are sometimes very injurious to flowers and fruit.

It has been said, however, that they also feed largely on plant-lice or aphides, which, if true, give them a very redeeming quality

The female earwig deposits her eggs under a stone in some cavity in the ground often dug out by her own labour.

Unlike most other insects, the female does not perish as soon as she has laid her eggs, but lives to behold her offspring, brooding over them almost like a hen. But it is distressing to learn that if the mother should die she is immediately devoured by her progeny.

The *Euplexoptera* are widely distributed over the surface of the earth. The tropical regions, however, can hardly claim any great predominance over more temperate climates as regards either the number or size of the species.

The largest European species (*Forficesita gigantea*), which is an inhabitant of some parts of England, measures about an inch in length of the body.

Our common Earwig (*Forficula auricularia*) is not only found all over Europe, but apparently throughout the greater part of the eastern hemisphere.

Fig. 77.—Forficula Auricularia.

Another British species, viz., the little Earwig (*Labia minor*), is also of wide distribution.

The earwig derives its name from its occasionally creeping into the human ear in search of concealment. This has been denied by many authors, but we have personally known instances of earwigs entering the ears of persons lying asleep in fields in the summer time. The insect can be driven out immediately by pouring a little oil into the ear.

Some writers have attempted to prove that the name earwig is a corruption of "earwing," in allusion to the shape of the hind wings; really a very ingenious idea, but incorrect nevertheless.

# ORDER HEMIPTERA.

INCLUDING THE LAND BUGS, WATER BUGS, PLANT BUGS, SKATERS, LANTERN FLIES, FROG HOPPERS, APHIDES, AND THEIR ALLIES.

The *Hempitera* are *Haustellate* insects; they have four wings, which are membranous and naked. Their metamorphosis is in-

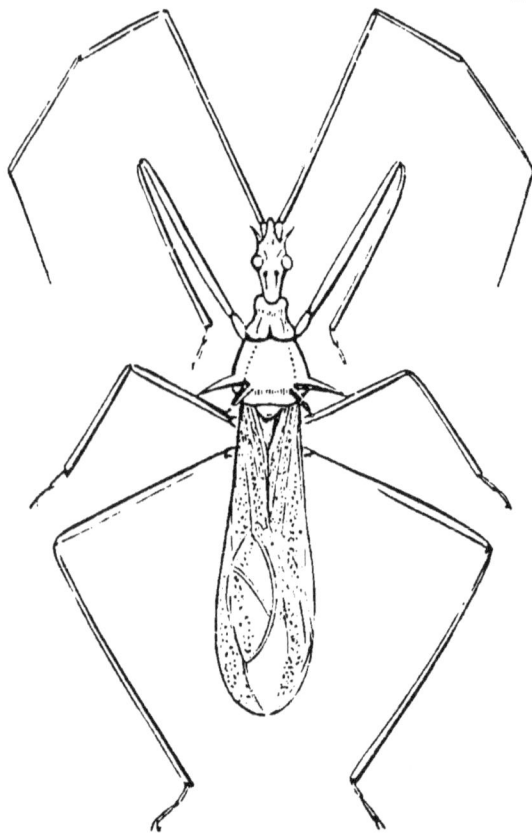

Fig. 78. – Zelus Quadrispinosus.

complete. The order is a very extensive one. The insects, however, are mostly found in the tropical parts of the world, from whence nearly twenty thousand species have been described. In the British Isles their number reaches nearly a thousand.

The study of the *Hemiptera* has hitherto been sadly neglected, owing to the great majority of them being very small species.

They include the Plant Bugs, Sand Bugs, Water Bugs, Cuckoo Spits, Froghoppers, Plant Lice, and many other familiar forms. The three great divisions into which they are divided are treated by many entomologists as distinct orders. The following will show

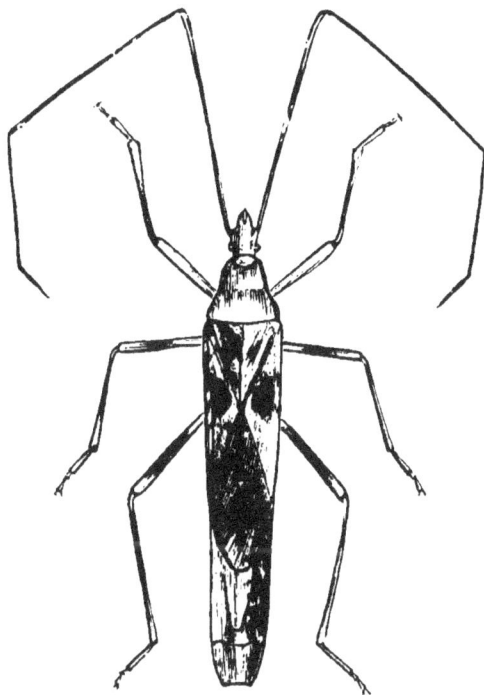

Fig. 79.—Macrocheraia Grandis.

under each heading the principal distinguishing character of each group or sub-order as we shall designate them.

    *Sub-Order* 1, *Heteroptera.*—Fore wings of a parchment like consistency.

    *Sub-Order* 2, *Homoptera.*—Fore wings membranous and naked, similar to the hind ones.

    *Sub-Order* 3, *Anaplura.*—Wingless; no metamorphosis.

The *Heteroptera* are regarded as the highest of the three groups, followed by the *Homoptera*, while the *Anoplura* constitute a

5

somewhat aberrant series, and are somewhat in the same position as the fleas (*Aphaniptera*) are to the true *Diptera*.

TABULAR VIEW

OF THE

PRINCIPAL FAMILIES OF THE HEMIPTERA.

*Sub-Order* 1.—*Heteroptera.*

TRIBE I.—GEOCORES OR LAND BUGS.

Family 1. Scutelleridæ or Shield Bugs.
,,     2. Coreidæ or Land Bugs.
,,     3. Lygæidæ or Chinch Bugs.
,,     4. Pyrrhocoridæ or Plant Bugs.
,,     5. Capsidæ or Sap Bugs.
,,     6. Tingididæ or Tree Bugs.
,,     7. Cimicidæ or Bed Bugs.
,,     8. Reduviidæ or Pirate Bugs.
,,     9. Emesidæ or Wolf Bugs.

TRIBE II.—HYDROCORES OR WATER BUGS.

Family 10. Saldidæ or Water Jumpers.
,,    11. Hydrometridæ or Skaters.
,,    12. Gerridæ or Ocean Bugs.
,,    13. Galgulidæ or Shark Bugs.
,,    14. Nepidæ or Water Scorpions.
,,    15. Notonectidæ or Water Boatmen.

*Sub-Order* 2.—*Homoptera.*

Family 1. Cicadidæ or Cicadas Flies.
,,     2. Fulgoridæ or Lantern Flies.
,,     3. Membracidæ or Horn Flies.
,,     4. Cercopidæ or Froghoppers.
,,     5. Tettigonidæ or Meadow Lice.
,,     6. Ledridæ or Oak Lice.
,,     7. Jassidæ or Elegant Lice.
,,     8. Psyllidæ or Plant Lice.
,,     9. Aphidæ or Aphides.
,,    10. Aleyrodidæ or Powder Flies
,,    11. Coccidæ or Scale Insects.

*Sub-Order* 3.—*Anoplura.*

Family 1. Pediculidæ or True Lice.

## SUB-ORDER I.—HETEROPTERA.

The *Heteroptera* constitute the first sub-order of the *Hemiptera*. They include the True Bugs, an extensive group of very varied structure and habits. The greater part of the terrestrial species feed on plants, though some feed on the blood of animals, while

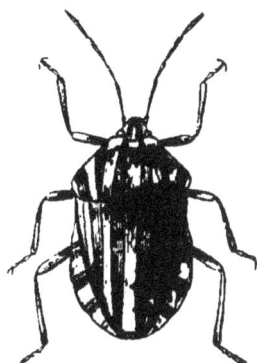

Fig. 80.—Graphosoma Lineatum (Mag.)

the aquatic species are principally carnivorous in their habits. This sub-order is divided into two tribes.

### TRIBE I.—GEOCORES OR LAND BUGS.

This tribe includes the terrestrial species of bugs. It is sub-divided into fifteen families.

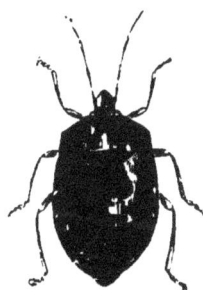

Fig. 81.—Tectocoris Banksii.

*Family* 1, *Scutelleridæ.*—This extensive family includes the greater portion of the largest and handsomest species of the Land Bugs.

The insects live upon plants, trees, and shrubs, and feed upon the juices, which they suck out of the soft tissues, many of them especially attacking juicy fruits. The family is very well represented in the warmer parts of the world, where also the most beautifully coloured species are met with.

One of the most handsome representatives in Europe is *Grapho-*

*soma lineatum,* which measures nearly half an inch in length.   It is reddish in colour and is common on flowers, especially those of the *Umbelliferæ,* but is not met with in our own country.

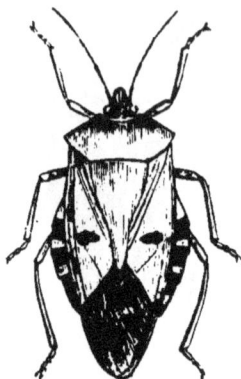

Fig. 82.—Catacanthus Incarnatus.

*Edessa cervus* is a native of South America, and another pretty species.

The most familiar British species is what is commonly known as the Colewort Bug (*Strachia oleracea*), a very pretty insect, which lives on cruciferous plants, and is said sometimes to be injurious in gardens.

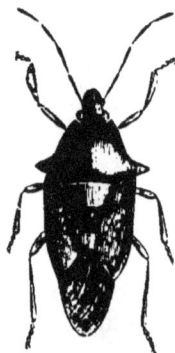

Fig. 83.—Euthyrhynchus Floridanus (Mag.)

Some of the Chinese Bugs belonging to the genus *Tesseratoma* are the largest species known.

*Family* 2, *Coreidæ.*—This family contains a great number of plant-feeding native and exotic species, varying considerably in shape and structure.   The majority of European species are, however, of small size in comparison to those found in tropical parts of the world.

The insects of this family are rarely adorned with bright colours, different shades of brown being the prevailing tints, although a few of them are gaily adorned.

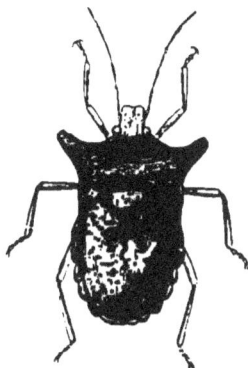

Fig. 84.—Trigonosoma Desfontainei.

In their general habits they much resemble the Shield Bugs.

Two British species are *Syromastes marginatus* and *Verlusia rhombea*.

*Family* 3, *Lygæidæ*.—The members of this family are, on the whole, much smaller than the *Coreidæ*, some of the smallest forms of which many of them resemble.

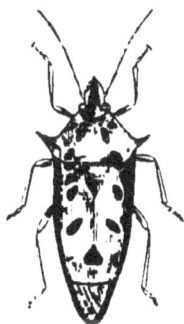

Fig. 85.—Cantao Ocellatus (Mag.)

These insects are generally of a red colour, with black bands and spots.

Several species are very injurious to cultivated plants. One of the most destructive of all is *Blissus leucopterus*, a black insect with white fore wings, each of which is marked with a large black triangular spot on the outer edge.  It measures about an eighth

Fig. 86.—Menenotus Lunatus.

of an inch in length.  The young larva is red.  In the United

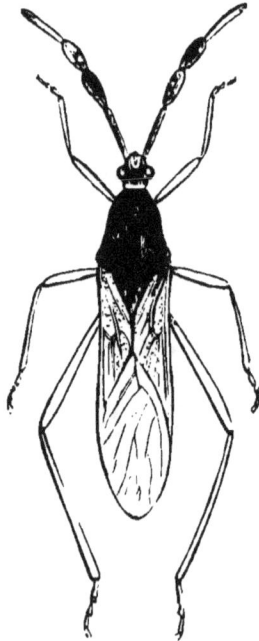

Fig. 87.—Copius Intermedius.

States this insect, which abounds to a considerable extent, is called the " Chinch Bug."

*Family* 4, *Pyrrhocoridæ.*—This family of bugs abounds in all parts of the world, and in Europe and Britain is undoubtedly the most numerously represented of all the families of bugs. A very common species in this country found on nettles is *Phytocoris tripustulatus,* which is about one-sixth of an inch.in length, and generally yellowish in colour.

*Family* 5, *Capsidæ.*—This family contains a great number of

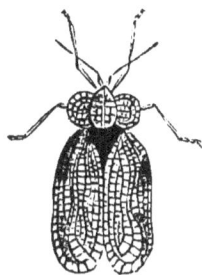

Fig. 88.—Tingis Pyri (Mag.)

small species of variegated colours, which feed exclusively on the sap of plants or the juice of fruits.

A rather small species (*Capsus ater*) is common upon herbage. The male is entirely black; in the female the head and thorax are reddish. The genus *Miris* and its allies include elongated species, which are found chiefly in grassy places.

*Family* 6, *Tingididæ.*—The species of this family differ con-

Fig. 89 —Cimex Lectularius (Mag.)

siderably among themselves in size, structure, and habits; and although the majority of them are carnivorous, others, especially among the smaller species, are herbivorous. One species (*Tingis pyri*), found in our own country, is sometimes very injurious to pear trees.

*Family* 7, *Cimicidæ.*—The type of this family is the common Bed Bug (*Acanthia lectularia*), which is only too well known to most people. Although treated as a British insect, it does not always appear to have been an inhabitant of these islands, but to have made its way here about the beginning of the sixteenth century. Three other British species have been described as inhabiting the dwelling-places of certain animals and sucking their blood. *A. columbaria* attacks pigeons. *A. hirundinis* is found in swallows' nests, and *A. pipistrelli* feeds on bats.

*Family* 8, *Reduviidæ.*—All the insects of this family are predaceous in their habits, and are exceedingly numerous in tropical climates. The largest British species is *Reduvius personatus*, an insect about three-quarters of an inch long, of a blackish-brown

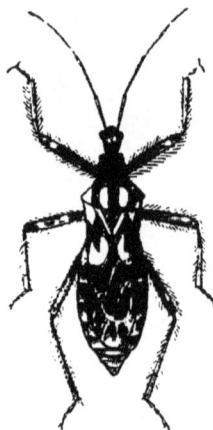

Fig. 90.—Acanthaspis Sexguttata

colour with reddish legs. It is well furnished with wings, and flies especially in warm summer evenings, when it frequently enters houses, being attracted by the lights. This insect is said to be a great enemy to the bed bug.

Some of the foreign species of *Reduviidæ* are most formidable insects, such as the great black *Conorrhinus renggeri* of Chili, which often attacks travellers when camping out.

*Family* 9, *Emesidæ.*—Most of these insects are carnivorous in their habits. The best known species of this family is probably *Plæaria vagabunda*, a brown, delicately-formed insect, which inhabits trees.

## TRIBE II.—HYDROCORES OR WATER BUGS.

All the insects belonging to this tribe are either water insects, or found only in the immediate neighbourhood of water. It contains the remaining families of the *Heteroptera*, six in number.

*Family* 10, *Saldidæ.*—The majority of species belonging to this family are small, dull-coloured insects, always found in the neighbourhood, though they are not strictly aquatic.

*Family* 11, *Hydrometridæ.*—The habits of most of these insects are herbivorous, and they are all found near water.

*Hydrometra stagnorum* is a black or brown insect, more or less tinged with reddish, and about half an inch long. It is found running on the surface of water, or else on the banks or among water plants, but is not so active as some of the other species.

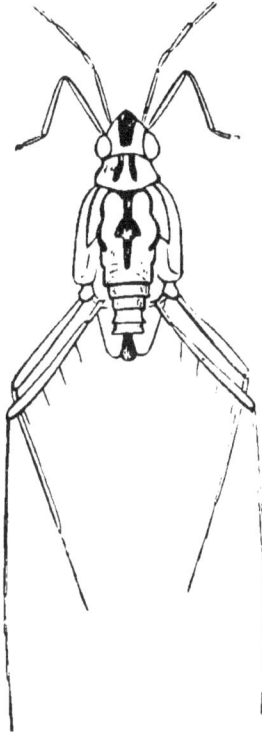

Fig. 91.—Halobates Pictus (Mag.)

*Family* 12, *Gerridæ.*—This family includes some very familiar insects, which may be seen running over the surface of every piece of water. These insects, of which several species are abundant in Britain, have boat-shaped bodies. The typical form is *Gerris lacustris*, which may be met with almost anywhere. They are predaceous in their habits, feeding upon other insects. Some nearly allied, but mostly very small species, with legs even longer in proportion than those of our common forms, are met with at sea within the tropics, and often at a great distance from land.

*Family* 13, *Galgulidæ.*—This family only includes a few American species, which are generally brown spotted with yellow. *Galgulus oculatus* inhabits the southern part of the United States, and measures about two-fifths of an inch in length.

*Family* 14, *Nepidæ.*—The species of this family are not very

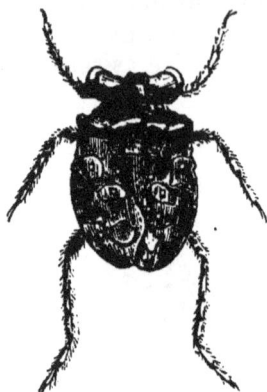

Fig. 92.—Galgulus Oculatus (Mag.)

numerous, but are of considerable interest. They are of large size, and very fierce and voracious. The genus *Betostoma*, found in the East Indies and America, includes the largest species of *Heteroptera*, some of which measure four and a half inches in

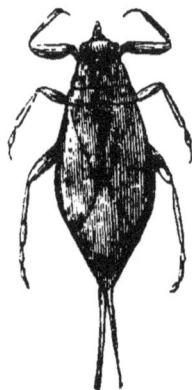

Fig. 93.—Nepa Cinerea.

length, and nearly six inches in expanse of the wing. Their food consists of small fish, frogs, etc.

The best-known species in our own country is the Water Scorpion (*Nepa cinerea*), which has the power of inflicting a very painful wound if handled. It measures about an inch in length.

In colour it is yellowish-grey, the back of the abdomen being red, and its large front legs, which somewhat resemble the claws of a

Fig. 94.—Notonecta Glauca.

scorpion, have given rise to the name by which it is popularly known. It is very common in stagnant water.

*Family* 15, *Notonectidæ.*—These insects are commonly known as " Water Boatmen," from their habit of rowing themselves

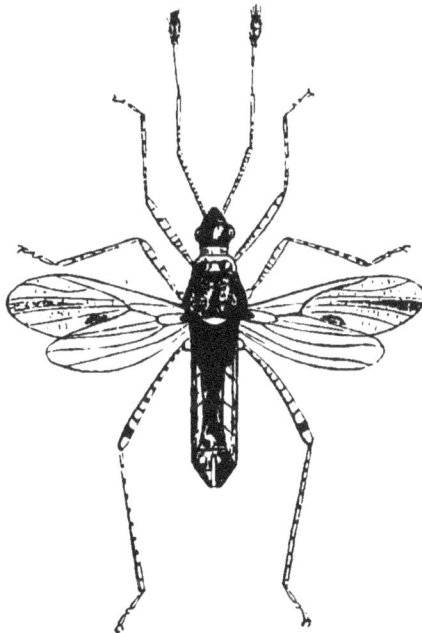

Fig. 95.—Metacanthus Punctipe (Mag.)

about on their backs with their long hind legs. They are car-nivorous, feeding on small insects and other " fry."

The most typical British species is *Notonecta glauca,* a yellowish insect about half an inch long. It is a most predaceous insect, and can bite severely.

Another common species inhabiting our own country is *Coriax*

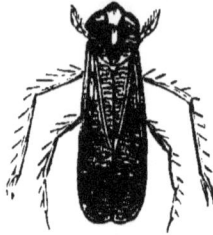

Fig. 96.—Corsa Striata.

*Geoffroyi*, which is nearly half an inch long. The *Notonectidæ* are all very powerful and active creatures.

## SUB-ORDER II.—HOMOPTERA.

This extensive sub-order includes the Cicadas, Lantern Flies, Plant Lice, Scale Insects, etc.

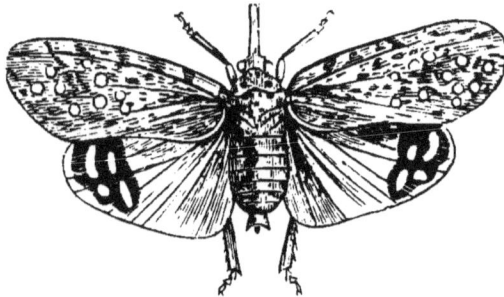

Fig. 97.—Fulogra Horsfieldi.

As already pointed out, the most striking general character of this group consists in the uniform texture of the forewings.

Fig. 98.—Cixius Cunicularius (Mag.)

There are eleven families included in the *Homoptera*, all the species belonging to which feed upon vegetable juices.

*Family* 1, *Cicadidæ.*—These insects are mostly inhabitants of warm climates, and our only British representative of the family (*C. anglica*) is one of the smaller species, the wings only expand-

Fig. 99.—Delphax Longipennis.

ing about an inch and a quarter. Four or five hundred species, however, are known to inhabit the world, and some of them attain to considerable dimensions.

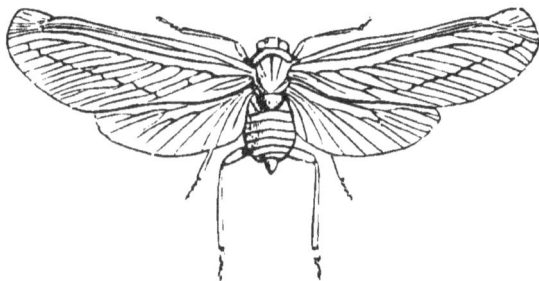

Fig. 100.—Derbe Strigipennis (Mag.)

They generally live upon trees and shrubs, and obtain their nourishment by piercing the tissues and sucking out the juices of their young tender shoots.

Fig. 101.—Issus Coleoptratus.

The *Cicadas* are improperly called "Locusts" both in America and Australia, on account of the chirping noise which they perpetually keep up in the woods, both day and night.

*Family* 2, *Fugloridæ.*—This family includes the Lantern Flies and Candle Flies, which are remarkable for their large size, bright colours, and strange forms.   They are almost all exotic insects.

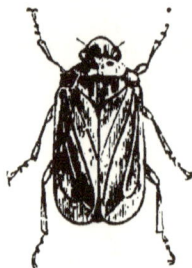

Fig. 102.—Tettigometra Virescens (Mag.)

The largest species is found in South America, and is called the Great Lantern Fly (*Fulgora la'ernaria*), measuring nearly three inches in length, and over four inches in expanse of the wings.

Fig. 103.—Pœciloptera Phalænoides (Mag.)

The Chinese Lantern Fly (*Fulgora candelaria*) has a red body, the fore wings being greenish with yellow spots, the hind wings orange colour with black tips.

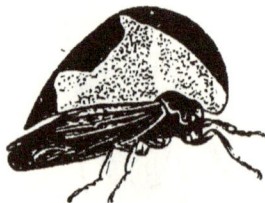

Fig. 104.—Membracis Foliata (Mag.)

*Cixius nervosus* is found in the British Isles, chiefly on alders. It measures about a quarter of an inch in length, and in colour is black with yellow legs, and transparent wings dotted with brown.

*Family* 3, *Membracidæ.*—This family is chiefly remarkable for the fantastic shapes assumed by the prothorax.

Most of the species are inhabitants of America, where they occur in wonderful abundance and variety.

A common species, *Centrotus cornutus*, is found in the British

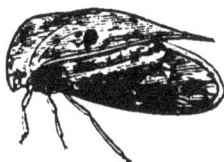

Fig. 105.—Smilia Fasciata (Mag.)

Isles and over the greater part of Europe. It is rather over a quarter of an inch long, black, with a pair of upright horns on the prothorax.

Another common European and British species (*Gargara gen-*

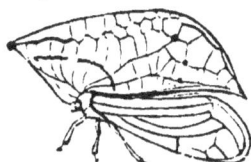

Fig. 106.—Œda Inflata Mag.)

*istæ*) is smaller than the preceding, and has no horns on the prothorax.

*Family 4, Cercopidæ.*—The " Froghoppers " and "Cuckoo-spits" belong to this family. They are chiefly small insects found among grass and bushes in the summer.

Fig. 107.—Physoplia Nigrata (Mag.)

A very common species in this country is the common Cuckoo-spit (*Aphrophora spumaria*), which is about a quarter of an inch in length, and of a yellowish-grey colour. This insect can make a prodigious leap in proportion to its size. It is said to sometimes spring to a distance of two yards. Its yellow *larvæ* may often be seen on grass and low plants enveloped in a mass of froth, which has given rise to the name of " Cuckoo-spit."

Fig. 108.—Hypsauchenia Westwoodii (Mag.)

*Family* 5, *Tettigonidæ.*—These insects are exceedingly numerous, and are often remarkably elegant in form. They are mostly inhabitants of America, where some three or four hundred species

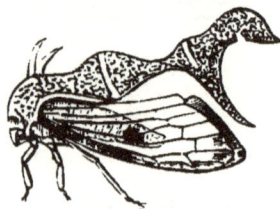

Fig. 109.—Heteronotus Vulneratus (Mag.)

have been described, but in England we have an exceedingly pretty species (*Tettigonia viridis*), which is common in damp meadows.

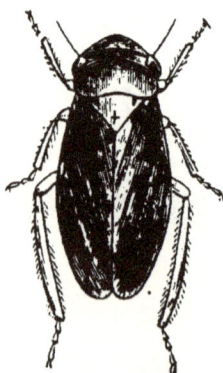

Fig. 110.—Jassus Atomarius (Mag.)

*Family* 6, *Ledridæ.*—Many of the insects belonging to this family resemble beetles somewhat in appearance. *Ledra aurita*, which may be considered the type of the family, is a greenish insect, about three-quarters of an inch in length, and found on oak trees.

*Family 7, Jassidæ.*—This family contains a considerable number of small insects, some of them of great beauty and elegant in appearance. The typical genus is *Jassus.*

*Family 8, Psyllidæ.*—This is the first family of the so-called " Plant Lice " or " Blight." They resemble the " Froghoppers "

Fig. 111.—Bythoscopus Venosus. (Mag.)

in their habits of jumping. Most of them are small insects, and they subsist on the sap of plants, to which they are sometimes injurious. A few species produce galls.

*Livia pyri* is a reddish insect, abundant on pear trees. Other common species are found on the alder, ash, oak, and nettle.

Fig. 112.—Livia Juncorum (Mag.)

*Family 9, Aphidæ.*—The *Aphidæ*, or " Plant Lice," " Blight," and " Green Fly," as they are variously called, are amongst the most destructive of insects. They are small and feeble creatures, but make up for their individual insignificance by their immense numbers.

They sometimes exist in countless millions, completely smothering the plants upon which they feed.

Migratory habits seem to be strongly developed among the *Aphidæ*, hence it is we are so suddenly visited by such enormous quantities of the insects during hot weather in the summer time.

In colour the *Aphides* are generally green, brown, and black, according to the species. Different kinds live on different plants and trees.

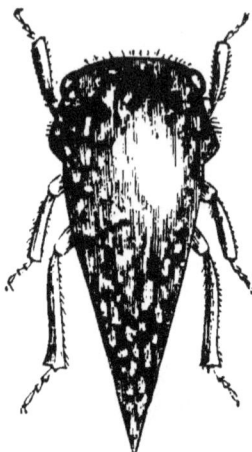

Fig. 113.—Darnis Limacodes (Mag.)

The Hop Aphis (*Aphis humuli*) lives upon the hop, and its

Fig. 114.—Tricophora Sanguinolenta (Mag.)

abundance or scarcity is a most important matter to the hop-growers in this country.

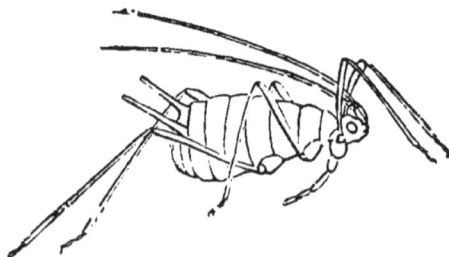

Fig. 115.—Aphis Tiliæ (Mag.)

One of the most destructive insects of the whole family is the Vine Aphis (*Phylloxera vastatrix*), which has committed terrible ravages at various times in most of the vine-growing districts on the Continent.

The Aphides have the habit of discharging a sweet sticky substance called honeydew, of which ants are very fond.

Fig. 116.—Lachnus Quescus (Mag.)

The reproduction of the *Aphidæ* constitutes one of the most interesting chapters in the history of the animal kingdom.

*Family* 10, *Aleyrodidæ.*—Many of the insects of this family resemble moths in appearance. Indeed, the typical species, *Aleyrodes*

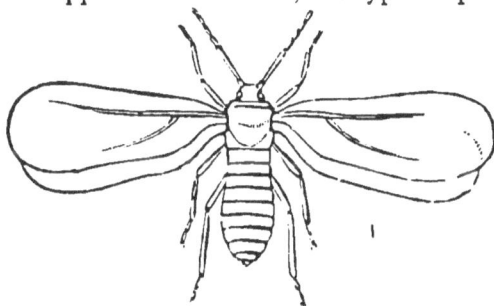

Fig. 117.—Aleyrodes Proletella (Mag.)

*proletella*, a very small reddish insect, was formerly considered to be a moth by the older writers on entomology.

*Family* 11, *Coccidæ.*—This family contains the Cochineal Insects, the Scale Insects, and their allies.

The *Coccidæ*, or Scale Insects, are sometimes very injurious to cultivated plants, but they are also useful, producing cochineal, shellac, manna, and other substances of considerable importance.

The Cochineal Insect (*Coccus cacti*) is a native of Mexico, and furnishes us with the most valuable and durable red dye that we possess, and the Lac Insect (*Coccus lacca*), an East Indian insect, produces the well-known lac-dye.

The common Scale Insect (*Coccus adonidum*) is well known in hothouses and conservatories in our own country, though it is not indigenous, having been imported from abroad.

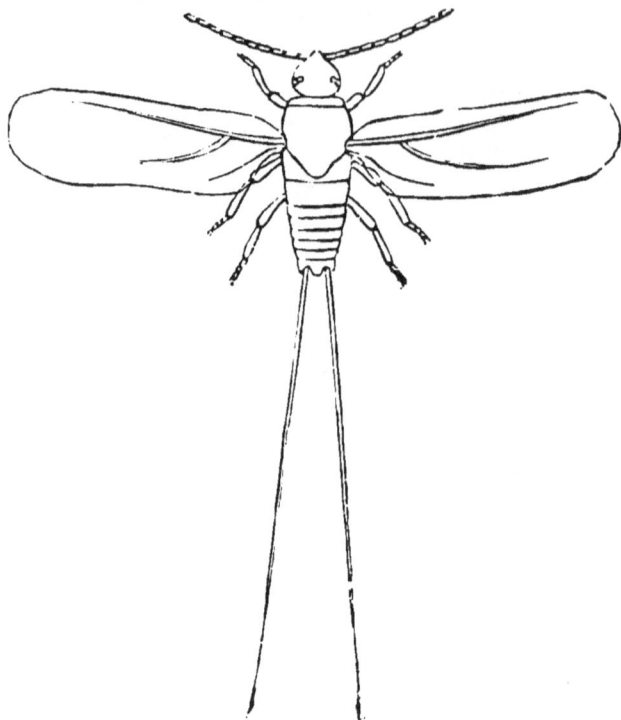

Fig 118.—Coccus Cacti (Mag.)

## SUB-ORDER III.—ANOPLURA.

This sub-order contains only one family, which is considered to be the last and lowest group of the *Hemiptera*, of which it may be

Fig 119.—Pediculus Capitis (Mag.

regarded as a very degenerate form.   The insects have no wings, and live upon the blood of other animals.

*Family* 1, *Pediculidæ.*—The " Lice " are a very extensive family, but at present not very well known. Almost every mammal has a louse peculiar to itself.

Three species infest man. The Head Louse (*Pediculus capitis*), found on the head, especially in children. The Body Louse *P.* (*Pediculus vestimenti*), found in the clothes; and the Crab Louse (*Phthirius inquinalis*), a broader and shorter insect, found in the hair on the face and body of uncleanly persons.

The lice infesting different races of men differ in colour—thus, those found on niggers are black; and those on Europeans are whitish.

Other species of lice infest the dog, cat, pig, mouse, elephant, monkey, etc.

The *Pediculidæ* undergo no metamorphoses.

# ORDER DIPTERA.

INCLUDING THE GNATS, MIDGES, CRANE FLIES, HAWK FLIES,
BEE FLIES, BREEZE FLIES, BOT FLIES, AND THEIR ALLIES.

---

The *Diptera*, or two-winged flies, are among the most numerous orders of insects, but they have been so little studied hitherto that we have but a very imperfect knowledge of them, especially as regards the tropical species.

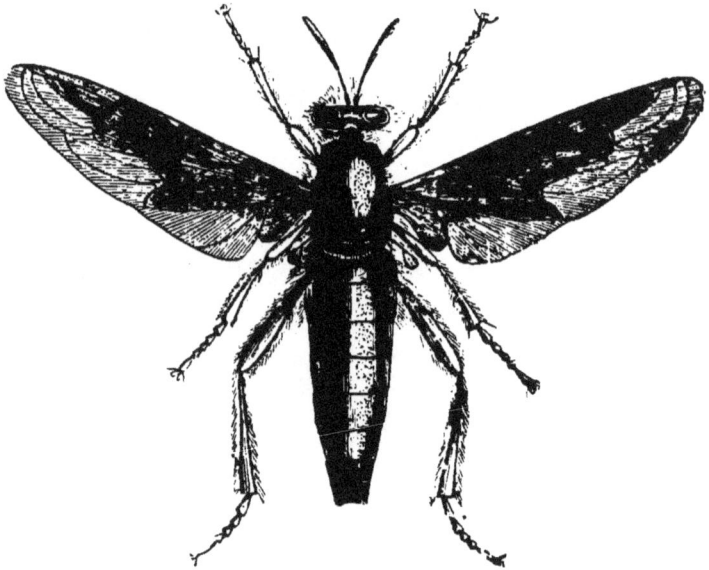

Fig 120.—Midas Giganteus.

The number recorded in the British Isles is about 3,000, that of the whole world probably exceeds 30,000, and it is certain that this number will be increased immensely as time goes on.

The *Diptera* may be defined as insects with a sucking mouth (*Haustellate*), and with only two wings, which are membranous, while the hind wings are represented only by a pair of small knotted organs, called *halteres*. Their metamorphoses are complete.

The *larvæ* are footless "grubs," generally with a soft body, but sometimes leathery or even nearly horny. Many of them possess a distinctly marked head, which may contain *ocelli*, but in the majority the head appears as if it were simply one of the neighbouring segments.

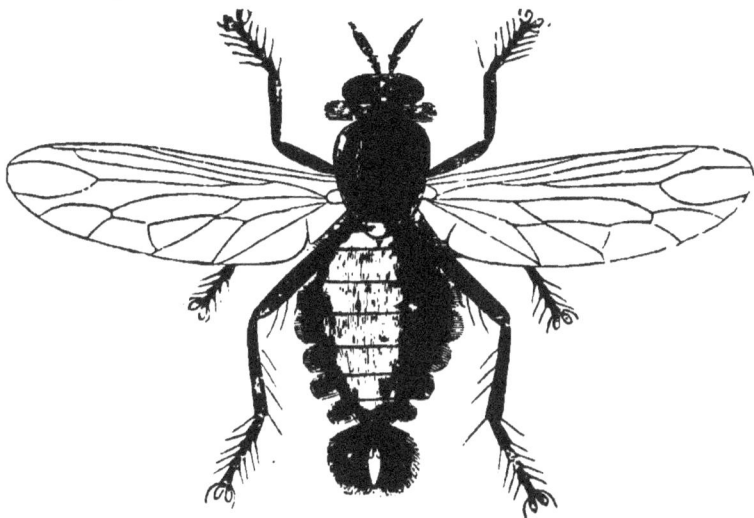

Fig. 121.—Craspedia Coriaria.

Their food generally consists of decayed animal and vegetable matter, and thus they are very useful as scavengers, particularly in hot climates.

The *larvæ* frequently live in the substance upon which they feed, while others reside in water, and many are parasitic.

For the classification of the *Diptera* we owe much to that eminent entomologist, Osten-Sacken, whose arrangement has been here generally adopted.

### TABULAR VIEW

#### OF THE

### PRINCIPAL FAMILIES OF THE DIPTERA.

#### TRIBE I.—NEMOCERA.

Family 1. Cecidomyiidæ or Wheat Midges.
,,   2. Mycetophilidæ or Army Worms.
,,   3. Simuliidæ or Sand Flies.

Family 4. Bibionidæ or Garden Flies.
,, 5. Blephariceridæ or Blood Flies.
,, 6. Culicidæ or Gnats.
,, 7. Chironomidæ or Midges.
,, 8. Orphnephilidæ or Bread Flies.
,, 9. Psychodidæ or Fungus Midges.
,, 10. Tipulidæ or Crane Flies.
,, 11. Dixidæ or Wood Gnats.
,, 12. Rhyphidæ or Dung Midges.

TRIBE II.—BRACHYCERA.

Family 13. Xylophagidæ or Tree Flies.
,, 14. Cœnomyiidæ or Poplar Flies.
,, 15. Stratiomyiidæ or Manure Flies.
,, 16. Acanthomeridæ or Pomade Flies.
,, 17. Tabanidæ or Breeze Flies.
,, 18. Leptidæ or Fox Flies.
,, 19. Asilidæ or Hawk Flies.
,, 20. Midaidæ or Wolf Flies.
,, 21. Nemestrinidæ or Flower Flies.
,, 22. Bombyliidæ or Bee Flies.
,, 23. Therevidæ or Hairy Flies.
,, 24. Scenopinidæ or Window Flies.
,, 25. Acroceridæ or Sloth Flies.
,, 26. Empidæ or Little Hawk Flies.
,, 27. Dolichopodidæ or Fairy Flies.
,, 28. Lonchopteridæ or Water Flies.
,, 29. Syrphidæ or Dart Flies.
,, 30. Conopidæ or Wasp Flies.
,, 31. Pipunculidæ or Hedge Flies.
,, 32. Platypezidæ or Fungus Flies.
,, 33. Œstridæ or Bat Flies.
,, 34. Muscidæ or Meat Flies.

Section 1.—Calypteræ.

Sub-family 1. Tachininæ or Parasitic Flies.
,, 2. Descinæ or Rainbow Flies.
,, 3. Sarcophaginæ or Screw Worms.
,, 4. Muscinæ or Blow Flies.
,, 5. Antromyiinæ or Summer Flies.

Section 2.—Acalypteræ.

Sub-family 6. Scatophaginæ or Dung Flies.
,, 7. Ortalinæ or Wood Flies.
,, 8. Trypetinæ or Fruit Flies.
,, 9. Piophilinæ or Cheese Flies.
,, 10. Diopsinæ or Horn Flies.
,, 11. Chloropinæ or Corn Flies.
,, 12. Drosophilinæ or Mould Flies.
,, 13. Agromyzinæ or Holly Flies.
Family 35. Phoridæ or Plant Flies.

TRIBE III. HOMALOPTERA.

Family 36. Hoppoboscidæ or Forest Flies.
  ,,   37. Nycteribidæ or Bat Lice.
  ,,   38. Braulidæ or Bee Lice.

TRIBE IV. APHANIPTERA.

Family 39. Pulicidæ or Fleas.

The number of families comprised in the *Diptera* is so great that space will only permit us to deal very briefly with each.

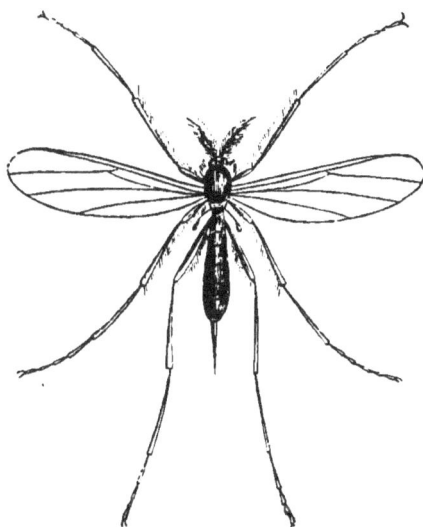

Fig. 122.—Epidosis Leucopeza (Mag.)

TRIBE I.—NEMOCERA.

This tribe contains many well-known insects—namely, the Midges, Gnats, Daddy-longlegs, etc., which are divided into twelve families.

*Family* 1, *Cecidomyiidæ.*—These are small delicate species, generally clothed with long hair. They are all vegetable feeders, some of them being very destructive to crops. The Hessian Fly (*Cecidomyia destructor*) is probably one of the best-known species, on account of the terrible mischief which it has caused in various countries, and particularly in the United States of America. Miss E. A. Ormerod, the well-known economic entomologist, has recently written a great deal concerning this insect, with the view to preventing its increase in our own country.

The Wheat Midge (*Diplosis tritici*) is another very mischievous species.

*Family* 2, *Mycetophilidæ.*—These are generally small species.

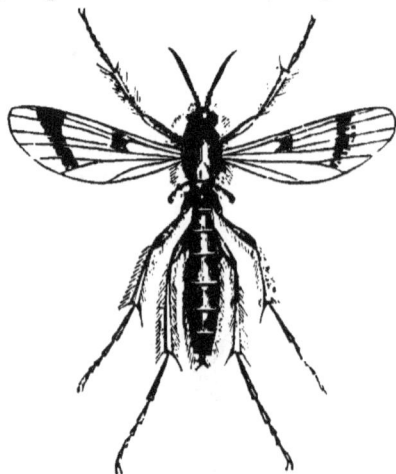

Fig. 123.—Mycetophila Distigma (Mag.)

They are all vegetable feeders. The *larvæ* live gregariously in fungi, rotten wood, bark, etc.

The Army Worms (*Sciara*) of America, which belong to this family, sometimes congregate in enormous numbers.

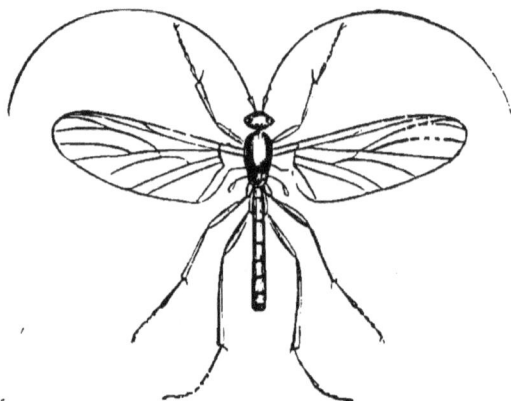

Fig. 124.—Macrocera Lutea (Mag.)

The flies are remarkable for their remarkable powers of leaping.

*Family* 3, *Simuliidæ.*—This family includes only one genus, *Simulium*.

It is, however, widely distributed, and some of the species, which are popularly called Sand Flies, are exceedingly annoying in hot climates.

*Family* 4, *Bibionidæ.*—Several species belonging to this group are very familiar insects. Among them is the St. Mark's Fly (*Bibio*

Fig. 125.—Simulium Ornatum (Mag.)

*Marci*), which is black, with transparent wings in the male and and brownish in the female.

These flies appear in the spring, and their *larvæ* live in dung or damp earth.

Many of the species are popularly called " garden flies."

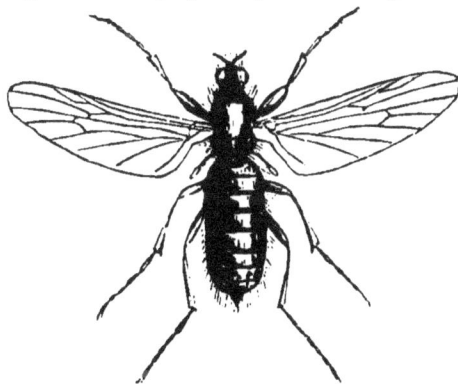

Fig. 126.—Bibio Hortulanus (Mag.)

*Family* 5, *Blephariceridæ.*—These flies have long legs, the *antennæ* are usually long and slender, and the wings broad and long. The type of this family is *Blepharicera fasciata*, the female of which is a very blood-thirsty creature. It is a dark brown fly, with transparent wings and yellow legs, and it measures about one-third of an inch in length.   It is, however, by no means abundant.

*Family* 6, *Culicidæ.*—The *Culicidæ*, or Gnats, are a very extensive family.

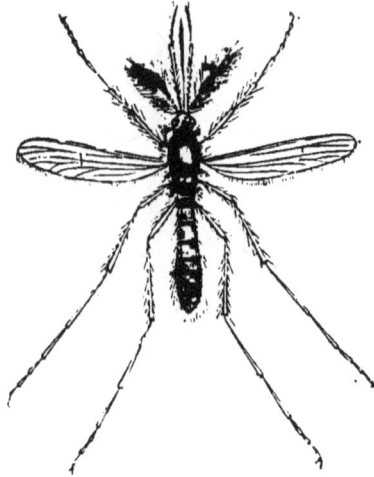

Fig. 127.—Culex Pipiens (Mag.)

The *larvæ* are nearly all aquatic.

They are generally small, dull-coloured insects, and are often called "Mosquitoes" in the country.   These insects frequently

Fig. 128.—Chironomus Plumosus (Mag.)

cause much annoyance during the summer months to persons engaged in out-door occupations.   They are particularly abundant in damp meadows, in the vicinity of water, and also in hay-fields, where the face and hands of the mowers are sometimes blistered to much pain by the suckers of the little creatures.

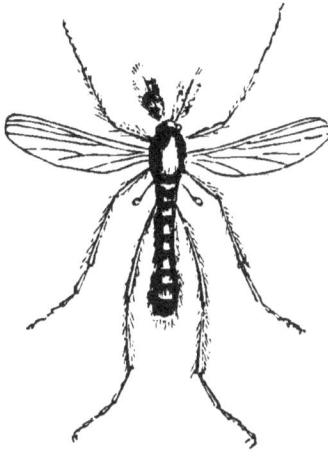

Fig. 129.—Ceratopogon Femoratus (Mag.)

One of the most troublesome species is probably the House Gnat (*Culex ciliaris*), which, thirsting for its evening meal, often

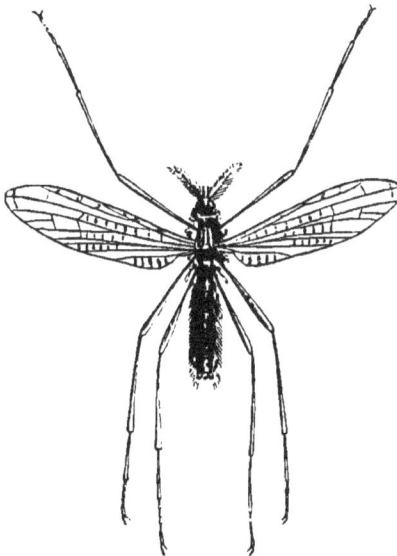

Fig. 130.—Trichocera Fuscata (Mag.)

enters our apartments, sounding its approach by a tolerably loud humming.

*Culex annulatus* is a little larger, and is said to produce a greater amount of irritation by its bite.

*Family* 7, *Chironomidæ.*—These insects are small delicate insects, much resembling gnats in appearance, and are popularly called "Midges." The *larvæ* of most of the species are aquatic.

The *larva* of *Chironomus plumosus*, which is common in

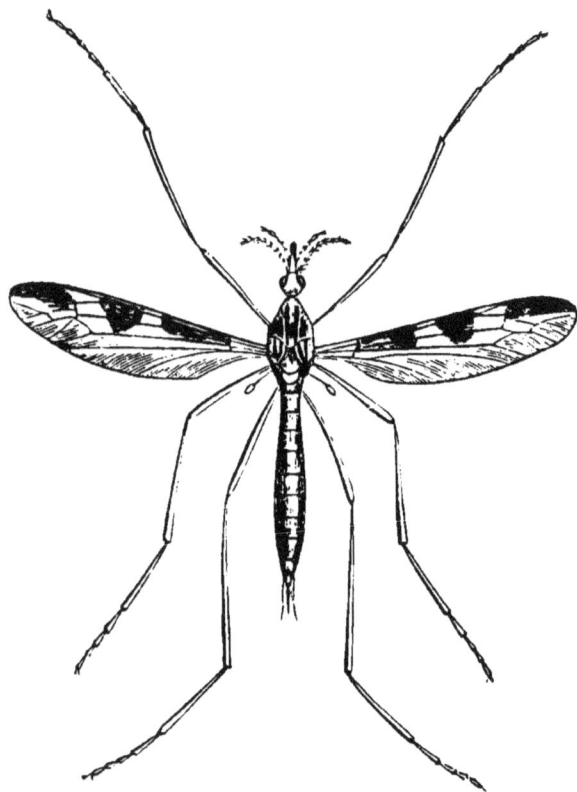

Fig. 131.—Tipula Gigantea.

stagnant water, is called the "blood worm" from its bright red colour.

*Family* 8, *Orphnephilidæ.*—Most of the species belonging to this family are of small size.

The typical species is *Orphnephelia testacea*, measuring scarcely one-tenth of an inch in length, and is sometimes found in bakehouses. It is of a rusty-yellow colour.

*Family* 9, *Psychodidæ.*—This is a small family consisting of minute brown or yellowish species remarkable for their resemblance to moths. They mostly feed on fungi and decayed vegetable substances.

The species of *Phlebotomus* are troublesome blood-suckers in the south of Europe.

*Family* 10, *Tipulidæ.*—The insects belonging to this group are popularly known as Crane Flies and Daddy-longlegs. They are very abundant.

The commonest species, *Tipula oleracea,* is a grey species with transparent wings, and exceedingly destructive to corn, roots of grass, etc., and no effectual remedy has hitherto been discovered for preventing their attacks.

The familiar Winter Gnat (*Trichocera hiemalis*) belongs to this

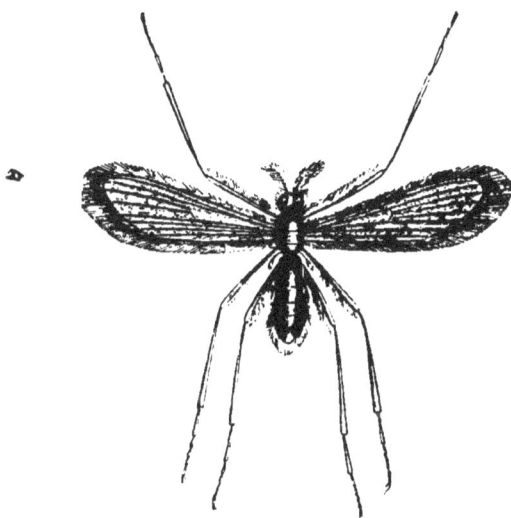

Fig. 132.—Erioptera Grisea (Mag.)

family. It may often be seen, even during the dreary months of December and January, dancing to and fro under the shelter of some hedge or wall. In frosty weather it conceals itself beneath the bark of trees, under leaves, etc.

*Family* 11, *Dixidæ.*—This family of gnats frequents damp places in woods, and are therefore designated " Wood Gnats." They may occasionally be seen during the summer time in immense swarms.

The species of *Dixa* are reddish, yellow, or black insects generally of small size.

*Family* 1, *Rhyphidæ.*—The *larvæ* of most of these insects feed upon decayed vegetable matter.

The species of *Rhyphus* may often be found resting on leaves in damp situations, they may also be seen on windows, and occasionally " dancing in the air."

They are generally brown, yellow, or grey insects of small size.

### TRIBE II.—BRACHYCERA.

These insects are generally of much larger size, and more robust in appearance than those contained in the previous tribe. They are divided into twenty-two families.

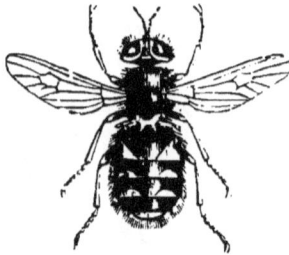

Fig. 133.—Stratiomys Chameleon.

*Family* 13, *Xylophagidæ.*—The *Xylophagidæ* are thick-looking insects. The head is as broad as the thorax. The legs, which are long and slender, are quite naked. The *larvæ* live in rotten wood, and the flies may often be seen at rest on the trunks of trees.

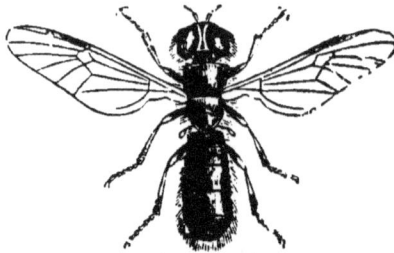

Fig. 134.—Sargus Cuprarius (Mag.)

*Family* 14, *Cænomyiidæ.*—In this family the head is narrower than the thorax ; otherwise they closely resemble the previous family. The typical European species is *Cænomyia ferruginea*, which measures about three-quarters of an inch in length, and in colour varies from rusty-yellow to black. The *larvæ* feed inside the trunks of rotten poplars.

*Family* 15, *Stratiomyiidæ.*—These are rather slender flies, varying from half to one inch in length. They are usually found resting on low plants. Many species frequent cow-dung and manure, etc.

Several kinds exhibit a metallic coloration, others are black, often more or less varied with white or yellow.

*Family* 16, *Acanthomeridæ.*—The *Acanthomeridæ* are met with principally in America. The typical species, *Acanthomera picta*, is found in Brazil, and measures more than an inch in length. It used formerly to be largely used in the manufacture of pomade.

*Family* 17. *Tabanidæ.*—The *Tabanidæ* are broad-looking

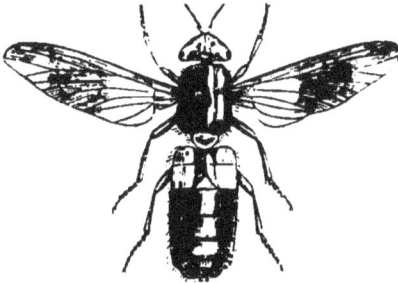

Fig. 135.—Chrysops Cæcutriens (Mag.)

insects. The *larvæ* usually live in damp earth. The females of these species live on the blood of animals.

The best-known species is the Gad Fly *Tabanus bovinus*), which measures more than an inch in length, and is particularly troublesome to cattle in hot weather. It is blackish above, and reddish beneath and on the sides of the abdomen.

The Clegg Fly (*Hæmatopota pluvialis*) also belongs to this family.

Fig. 136.—Hæmatopota Pluvialis (Mag.)

It is a dingy-looking insect with mottled-grey wings, and is some·times very abundant in damp meadows.

Many other species of *Tabanidæ*, popularly called "Breeze Flies," are plentiful in this country.

*Family* 18, *Leptidæ.*—These insects are mostly predatory, though many small species are parasitic on animals and other insects. The *larva* of *Vermilio degeerii* lives in sand, and feeds on such insects as fall in its way.  Those of the genus *Leptis* are parasitic on cockchafers and other large beetles.

*Family* 19, *Asilidæ.*—The *Asilidæ* or Hawk Flies are a family of considerable extent, and include many very large and conspicuous insects.  They feed on other insects, and are very courageous, sometimes attacking even dragon flies.  One of the most common

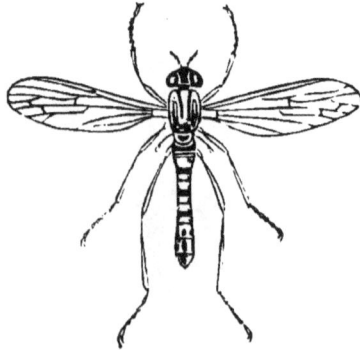

Fig. 137.—Vermilio Degeerii (Mag.)

species is *Asilus crabroniformis,* which is common in the south of England.

*Family* 20, *Midaidæ.*—These are large insects resembling the *Asilidæ* in their habits, and are nearly all tropical.  A few species, however, inhabit the south of Europe.

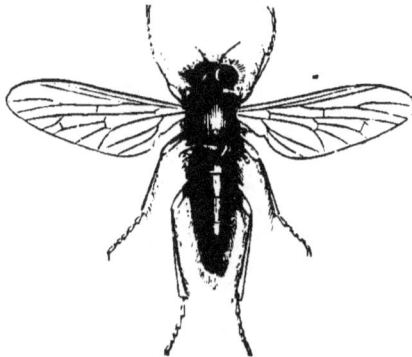

Fig. 138.—Chrysopila Atrata (Mag.)

*Midas giganteus* is found in Brazil and adjacent countries in tropical America.  It measures about one inch and a half in length.

*Family* 21, *Nemestrinidæ.*—These insects are chiefly inhabitants of tropical climates like the preceding family.  They are generally of a black colour, sometimes with red legs.  They suck the nectar of flowers through a long proboscis, but nothing is yet known of their transformations.

*Family 22, Bombyliidæ.*—The Bee Flies belong to this family. They are small or middle-sized insects, densely clothed with hairs. They are exceedingly active insects with long tongues, with which

Fig. 139.—Xylophagus Ater (Mag.)

they suck honey from flowers. Their *larvæ* are often parasitic in their earlier stages on other insects.

They are very useful in North America by keeping down the number of locusts, in the egg capsules of which the *larvæ* feed.

Two species, *Bombyliis major* and *B. medius*, are common in this

Fig. 140 —Ceria Conopsoides (Mag.)

country, and may often be seen in gardens and open spaces in woods during the spring and summer months.

*Family 23, Therevidæ.*—This family is a very numerous one, especially those of the typical genus *Thereva*. They are slender, middle-sized black insects, and their bodies are generally covered with hair. They are generally found in the neighbourhood of bushes and trees.

*Family* 24, *Scenopinidæ.*—The *Scenopinidæ* include a few minute, black flies, not exceeding a quarter of an inch in length. They are not very common, but are sometimes found about windows.

*Family* 25, *Scroceridæ.*—This family also contains small species of flies, but their metamorphoses are at present unknown.

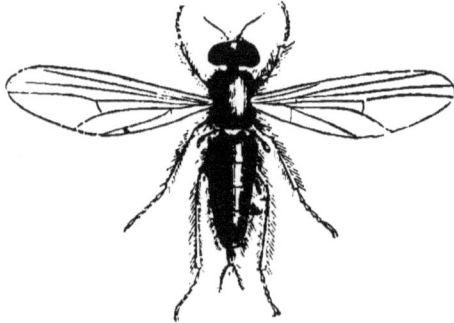

Fig. 141.—Medeterus Nolatus (Mag.)

They are usually to be found resting easily on dry branches of trees, but are generally rare.

*Family* 26, *Empididæ.*—The insects of this group are very

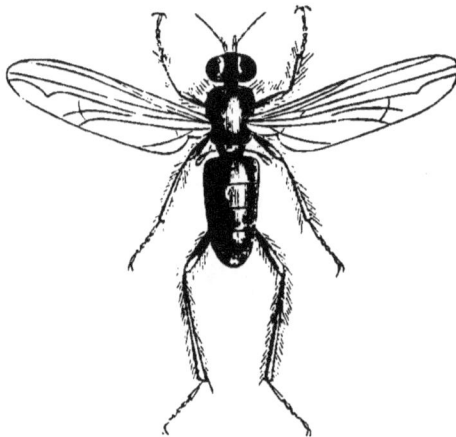

Fig. 142.—Dolichopus Regalis (Mag.)

similar to the *Asilidæ* in their carnivorous propensities, but the species are usually of a much smaller size. *Empis livida* is a common yellowish species measuring about one-third of an inch in length.

*Family* 27, *Dolichopodidæ.*—This family is a very extensive one, including a considerable number of genera and species. They are usually small flies often of brilliant metallic colouring. They may be found generally in the neighbourhood of water.

*Family* 28, *Lonchopteridæ.*—The species of *Lonchopteridæ* are small yellowish-brown or greenish flies with lanceolate wings.

Fig. 143.—Scenopinus Fenestralis (Mag.)

They frequent the banks of shady streams, and are often found resting on wet stones.

*Family* 29, *Syrphidæ.*—The majority of flies belonging to this family are of a moderate size, and possessed of brilliant colours. With the exception of the enormous family *Muscidæ*, they are one of the largest families of the *Diptera*.

Most of the species may be recognised at once by their pecu-

Fig. 144.—Syrphus Ribesii (Mag.)

liar mode of flight, for they hover motionless in the air, and if alarmed dart off with a rapid motion which the eye cannot follow.

*Milesia Crabroniformis* is a very large species common in the south of Europe.

About thirty species are found in the British Isles. Of these *Syrphus pyrastri* is perhaps the best-known example. It is a very wasp-like creature, and is, indeed, mistaken by many people for a wasp. It is a very useful insect, feeding largely on aphides and plant lice, and should therefore be encouraged by every horticulturist.

*Family* 30, *Conopidæ.*—The *larvæ* of most of these insects are parasitic on *Orthoptera* and *Hymenoptera.* Many species have a resemblance to wasps in appearance, being striped with yellow and black. The typical genus, *Conops*, are slender flies measuring half an inch in length.

*Family* 31, *Pipunculidæ.*—These are black or brown insects. The head is large and round, generally broader than the thorax. The *larvæ* are mostly parasitic on other insects. The flies may often be seen in great swarms in shady places in the vicinity of hedges, in lanes, etc.

*Family* 32, *Platypezidæ.*—This family contains many beautiful

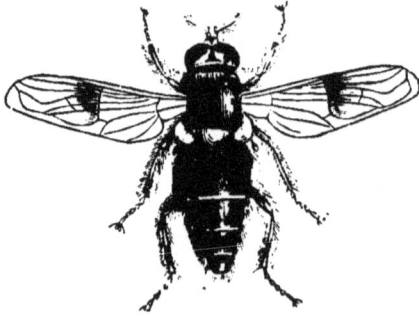

Fig. 145.—Eristalis Simili (Mag.)

flies. Great numbers of them are marked with orange and silver and black. The *larvæ* inhabit fungi.

*Family* 33, *Æstridæ.*—The Bot Flies are well known on account of the annoyance they cause to cattle. They lay their eggs upon the hides of the animals, and the *larvæ*, when hatched, make their way under the skin, and there take up their abode, living on the juices of their unfortunate host. It is estimated that the loss occasioned by these troublesome pests amounts to millions of pounds in the British Isles alone. The best-known species is probably *Æstrus bovis*, which infests the ox. The cattle are so well aware of the danger attending the presence of this insect, that as soon as it appears near them, the whole herd exhibits the most unmistakable signs of terror, rushing about their pasture with their tails in the air, and in case of need taking refuge in water, where the fly will not follow them. Miss E. A. Ormerod has recently paid much attention to the economy of this insect, with the view to finding some remedy for its attacks.

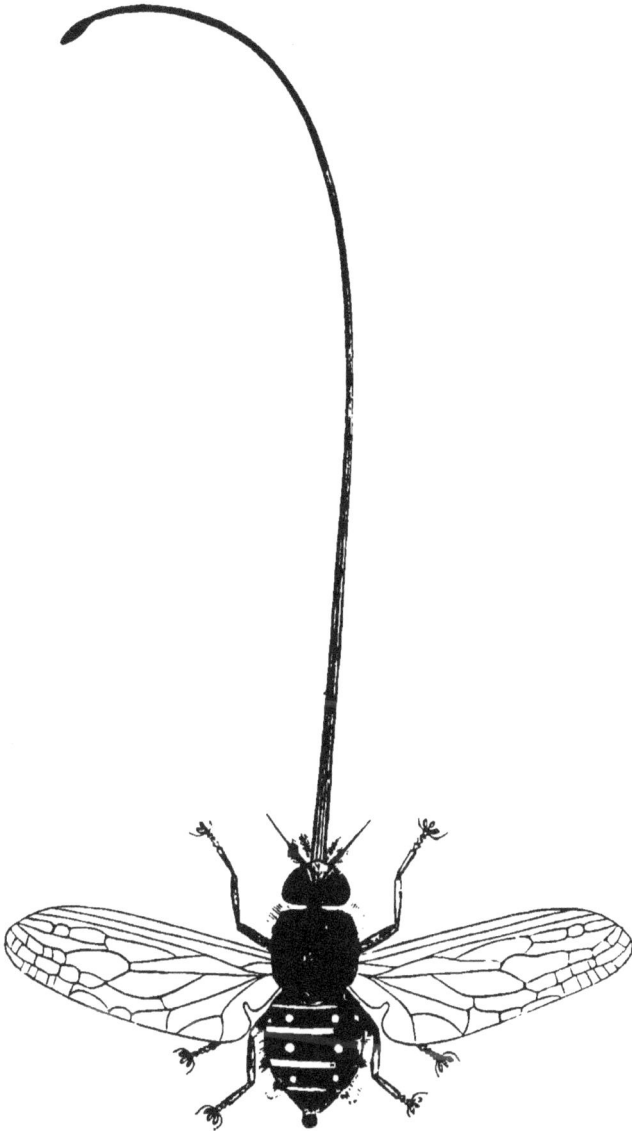

Fig. 145.—Nemestrina Longirostris (Mag.)

*Family* 34, *Muscidæ.*—The *Muscidæ* is the most numerous family, of the *Diptera* and it includes many familiar insects. It is divided into two sections, containing thirteen sub-families. The first section, *Calypteræ*, contains five sub-families. The *Tachininæ*

Fig. 147.—Volucella Zonaria.

are parasitic in the larval stage on *Lepidoptera*. The largest of the *Muscidæ* found in Europe is *Echinomyria grossa*, which measures about three-quarters of an inch in length.

Fig. 148.—Conops Flavipes (Mag.)

Many species of the *Dexinæ* are of the most metallic-green or blue. They are principally found in Australia.

The Flesh Fly (*Sarcophaga carnaria*) belongs to the *Sarcophaginæ*

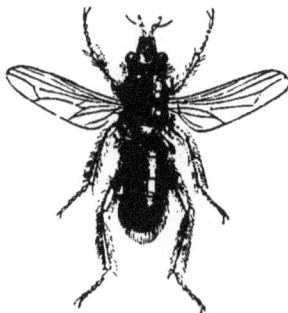

Fig. 149.—Myopa Ferruginea (Mag.)

and is one of our largest species. The *larvæ* of some species are called "Screw Worms" in America.

The fourth sub-family, the *Muscinæ*, contains the most typical species of the *Diptera* or two-winged flies. The common House Fly (*Musca domestica*) is the most familiar example. Several blue-and-green flies are produced from *larvæ* which feed on more or less putrid flesh, and are well known to anglers as "Gentles." The Blow Fly or Blue Bottle (*Musca vomitoria*) is one of these,

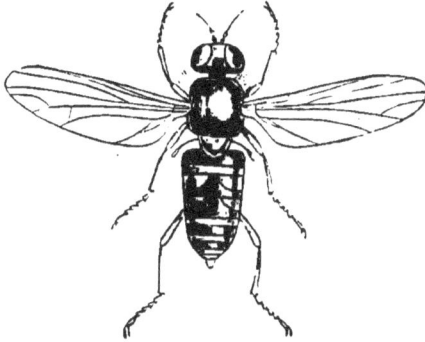

Fig. 150.—Platypeza Fasciata (Mag.)

also the Green Bottle Fly (*Lucilia Cæsar*) which is often common about hedges in the summer time. Many species belong to the sub-family *Anthomyiinæ*, are very small insects, and their *larvæ* mine in the leaves of plants like those of *Tineæ*

Section 2, namely *Acalypteræ*, contains eight sub-families.

The species of *Scalophaginæ* are usually called "Dung-flies."

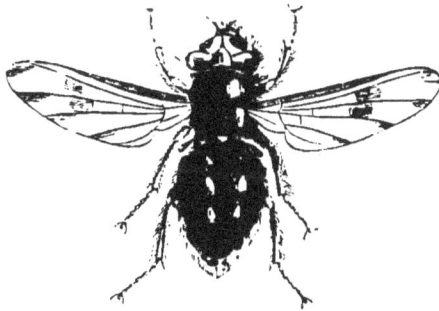

Fig. 151.—Gastrophilus Equi (Mag.)

The commonest species, the Yellow Dung Fly (*Scatophaga stercoraria*), feeds largely on manure and other refuse.

The *Ortalinæ* are flies of small or moderate size, which are met with principally in woods and fields. The *Trypetinæ* feed chiefly on fruit. One species, *Ceratitis hispanica*, is very destructive to oranges in the south of Europe, etc Of the *Piophilinæ*, *Piophila-casei* may be taken as the typical example. The *larvæ* which abound in cheese and bacon are popularly called "Cheese Hoppers," or "Jumpers," in allusion to their habit of springing. The *Diopsinæ* are principally met with in the tropical parts of the world. They are generally insects of moderate size. Most of the

Fig. 152.—Hypoderma Bovis (Mag.)

species of *Chloropinæ* are of small size. Their *larvæ* generally live in the stems of corn and various kinds of grasses. The *Drosophilinæ* are small, dull-coloured flies, the *larvæ* of which feed on fungi, rotten fruit, and similar substances.

The last sub-family, namely, the *Agromyzinæ*, is very extensive. One of the commonest species is *Phytomyza ilicis*, the *larva* of which forms large brown blotches on the leaves of the holly.

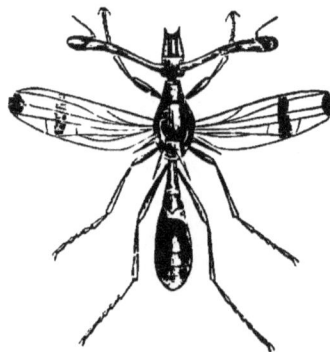

Fig. 153.—Diopsis Subfasciata (Mag.)

*Family* 35, *Phoridæ.*—The insects belonging to this family are generally of small or moderate size. They are very active, and may be observed on plants, and also sometimes sunning themselves on windows.

### TRIBE III.—HOMALOPTERA.

The insects belonging to this section much resemble spiders in appearance. They are all parasitic, and are remarkable for the perfect insect producing its young singly, and that not in the egg state, as in most insects, but either in the *pupa* state, or as a mature *larva* ready to become a *pupa* immediately. This tribe contains only three families.

Family 36, *Hippoboscidæ.*—The most familiar insects comprised in this family are known as "Forest Flies." The common Forest Fly (*Hippobosca equina*) is a brown species about one-third of an inch in length, and exceedingly annoying to horses, particularly in

Fig. 154.—Oscinis Cornuta (Mag.)

the New Forest. The Sheep Tick (*Melophagus ovinus*) is another well-known example.

Family 37, *Nycteribidæ.*—These insects are all parasitic on bats, and are commonly known as "Bat Lice." They are wingless, but have a pair of halteres placed upon the dorsal surface.

Family 38, *Braulidæ.*—This family includes only a single species,

Fig. 155.—Phora Camariana (Mag.)

which is parasitic upon the honey bee. Its name is *Braula cæca,* and it is an exceedingly minute creature.


### TRIBE VI.—APHANIPTERA.

The tribe *Aphaniptera* was formerly considered to be a separate order, but it is now generally included among the *Diptera* on account of the similarity in the transformations besides other considerations. There is only one family.

*Family* 39, *Pulicidæ.*—To this family belongs the numerous species of fleas which infest different kinds of animals.

The Human Flea (*Pulex irritans*) is familiar to every one. It gets its living by sucking our blood, and is of great annoyance, particularly in hot weather. The blisters which it raises on various parts of the body are exceedingly irritating.

The Dog Flea (*Pulex canis*) and the Cat Flea (*Pulex felis*) are

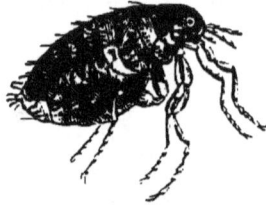

Fig. 156.—Pulex Irritans (Mag.)

distinct from the above, and are chiefly confined to those animals, although they do not disdain to vary their diet occasionally with human blood.

The Jigger or Chigoe (*Sarcopsylla penetrans*) is abundant in the West Indies and in South America. It is sometimes very annoying to travellers, burrowing deeply under the skin. The body of the female is capable of attaining to the dimensions of a pea.

www.ingramcontent.com/pod-product-compliance
Lightning Source LLC
Chambersburg PA
CBHW021944190326
41519CB00009B/1138